D1450955

The Seven Day Circle

The Seven Day Circle

The History and Meaning of the Week

Eviatar Zerubavel

The University of Chicago Press

Chicago and London

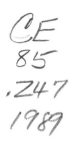

The University of Chicago Press, Chicago 60637
The University of Chicago Press, Ltd., London

98 97 96 95 94 93 92 91 90 89   5 4 3 2 1

The edition is reprinted by arrangement with
The Free Press, a division of Macmillan, Inc.

Library of Congress Cataloging in Publication Data

Zerubavel, Eviatar.
  The seven day circle : the history and meaning of the week /
Eviatar Zerubavel.
        p.   cm.
  Reprint. Originally published: New York : Free Press ; London :
Collier Macmillan, c1985.
  Bibliography: p.
  Includes indexes.
  1. Week—History. 2. Week—Social aspects. I. Title.
[CE85.Z47 1989]
529′.1—dc19   88-30393
ISBN 0-226-98165-7 (pbk.; alk. paper)   CIP

⊗ The paper used in this publication meets the minimum
requirements of the American National Standard for Information Sciences—Permanence of Paper for Printed Library Mate-
rials, ANSI Z39.48-1984.

To my mother,
  whose life is marked by Thursdays

Imagine for a moment that the week suddenly disappeared. What a havoc would be created in our time organization, in our behavior, in the co-ordination and synchronization of collective activities and social life, and especially in our time apprehension. Many of us would certainly mix our appointments, shift and change our activities, and fail many times to fulfill our engagements. If there were neither the names of the days nor weeks, we would be liable to be lost in an endless series of days—as gray as fog—and confuse one day with another. We think in week units; we apprehend time in week units; we localize the events and activities in week units; we co-ordinate our behavior according to the "week"; we live and feel and plan and wish in "week" terms. It is one of the most important points of our "orientation" in time and social reality.

Pitirim A. Sorokin,
Sociocultural Causality, Space, Time

Contents

List of Figures

Acknowledgments

Three sociologists in particular influenced the way I have come to think about time—Henri Hubert, Emile Durkheim, and Pitirim Sorokin. Three additional sociologists, whom I have been fortunate to know in person, indirectly influenced my decision to study the week. The late Erving Goffman taught me how to focus on those ubiquitous aspects of everyday life that are normally taken for granted and therefore overlooked. Peter Berger inspired my interest in further clarifying the fundamental distinction between natural and social reality. Renée Fox helped me become more aware of the tremendous significance of that mysterious entity we call "culture."

I am particularly indebted to Lynn Collins, who was extremely helpful throughout the past two and a half years, and especially at the stage when my ideas about the week were just beginning to crystalize. I would also like to thank Barry Schwartz, Deborah Wolfe, Rogers Brubaker, Laurie Beck, Julie Gricar, Dean Savage, Charles Lidz, Ronald Burt, Linda John, Dudley Duncan, and my father Berl Zerubavel for reading an early draft of the manuscript and providing me with solid advice which helped me strengthen my arguments, abandon dead-end tracks, and discover new paths. They, as well as Allan Silver, Joshua Lederberg, Laura Downs, Jonathan Imber, David Crook, and David Powell, also directed me to some particularly rich sources of information. Emanuel Smikun, Gisela Joppich, and Lynn Rapaport provided me with translations of some very useful material in Russian and German.

My understanding of the way modern humans live with the week and experience it has been furthered by Ron Hiram, Jack Kugelmass, Tamar Ariav, Yaakov Reshef, Manette Berlinger, Batya Perlson, Gad Ariav, Roberta Lentz, and Stephanie Hiram, who generously shared with me the known and less-known details of their weekly schedules and routines.

I would also like to express my gratitude to my Free Press editor, Joyce Seltzer, for helping a writer in his early efforts to break away from the habit of writing exclusively to an academically oriented audience.

My daughter, Noga Zerubavel, played a central role in inspiring me to write this book. Her inquisitive mind and intellectual curiosity also led me to address various weekly patterns which I had previously taken for granted and thus overlooked. My wife, Yael Zerubavel, helped me edit the final draft of the manuscript and was nearly always willing to hear me read to her "just one more paragraph." For the tremendous moral encouragement and support with which she provided me through her faith in my writing, particularly during the most difficult period in my professional career, I shall always remain grateful.

September 1984
New York

Introduction:
"Daddy, What's Thursday?"

O NE SUMMER AFTERNOON several years ago, as we were heading back home from the park, I mentioned to my 3½-year-old daughter something I planned for the following Thursday, when my inquisitive young friend asked me: "Daddy, what's Thursday?" As I soon found out, that was much easier to ask than answer, and to this day I still regret not having taped my desperate attempts to satisfy her refreshing and inspiring curiosity. Yet it was on that afternoon that the seeds of a book on the week were sown.

Four years have passed since then, and the questions—my own, now—have only become more numerous and more intriguing. Where did the seven-day week come from? Why was it invented? Have there been other weekly cycles of a different length? Why are many radio programs aired regularly every seven days rather than every four or every nine days? Why are suicide rates highest on Mondays and lowest on weekends? Why is it particularly difficult to keep to a strict diet on weekends? Why are we often unproductive on Friday afternoons? Why is it more common to visit one's family every other Sunday than, say, every sixteen days? Why is it easier to mistake a Wednesday for a Thursday than a Sunday for a Monday? Why does the Wednesday immediately preceding Thanksgiving Day often "feel" like a Friday and the Tuesday immediately following Labor Day like a Monday? Why do Wednesday night dates "count" less than Saturday

1

night ones? Why is Saturday night the time we feel the loneliest when staying at home by ourselves?

The week occupies an important place in our minds. The fact that, as children, we learn about "the weekend" long before we become acquainted with either "June" or "the 14th" indicates that the weekly organization of our environment may be far more salient to us than either its annual or monthly structure. When making plans for a particular date, we usually first check on what day it falls, and stadiums and theaters usually advertise their attractions as scheduled for, say, "Monday, August 11," despite the fact that "August 11" by itself would have been just as precise.

Much of our social environment is structured along weekly patterns (to the point where we often can tell what day it is merely by looking around us). In order to navigate successfully within society, we require a sort of mental "temporal map" that informs us, for example, that the best day for spending a relaxed morning with our parents is Sunday, that museums are often closed on Mondays, and that there are reduced rates for long-distance telephone calls on weekends. On the basis of such a "map," we may avoid Saturday nights at movie theaters, Fridays at banks, and Saturday afternoons at supermarkets and department stores. Such a "map" also serves to remind me that, if I wish to have a long telephone conversation with a particular friend, I should avoid calling him on Wednesday evenings, when he regularly watches his favorite television show.

The effective use of such a "temporal map" obviously presupposes our keeping track of the days of the week. Missing the regular weekly business meeting, class, or clinical appointment is one of the consequences of an inability to recall what day it is. Such an inability is much less common than the inability to recall today's date and is one of the major manifestations of the pathological mental state of "disorientation."[1] Thus, for example, when Leo Tolstoy's Ivan Ilych does not seem to care whether it is Friday or Sunday,[2] it is quite obvious that he is dying, since the living would hardly ever risk ignoring the week.

Recalling what day today is is one of the first things we usually do upon waking, since it is indispensable for transcending our subjectivity and participating—at least mentally—in a social, rather than a merely personal, world.[3] The uneasy feeling that accompanies our realization that we have lost count of the days of the week[4] is essentially the well-justified anxiety about being barred from full participation in our social environment. In other words, adhering to the week protects us from the dreadful prospect of practical exile from the social world.

This explains why, when finally being freed after a prolonged period of social isolation as a hostage of the Italian Red Brigades,

General Dozier immediately asked what day it was.[5] That adhering to the seven-day cycle is an integral part of people's effort to remain "civilized" even when away from society is also nicely demonstrated in Daniel Defoe's classic novel about a castaway sailor:

> After I had been there about ten or twelve days, it came into my thoughts, that I should lose my reckoning of time for want of books and pen and ink, and should even forget the Sabbath days from the working days; but to prevent this I cut it with my knife upon a large post. . . . Upon the sides of this square post I cut every day a notch with my knife, and every seventh notch was as long again as the rest.[6]

(It is by no means a mere coincidence that when, following a long period of total social isolation, Robinson Crusoe finally encounters another human being, he actually names him after the day of the week on which he meets him, Friday!) Humans' attempt to cling desperately to a social world to which they once belonged, through adherence to the seven-day rhythm, is also well portrayed in Dalton Trumbo's novel *Johnny Got His Gun*. Despite being fully aware of the arbitrary—and most likely erroneous—basis of the way he anchors his weekly cycle in historical time, the protagonist nevertheless finds it crucial to fill up the universe he creates in his mind with familiar temporal territories such as Sundays and Mondays.[7] (Quite similarly, nearly two thousand years ago, Jewish sages recommended that travelers who lose count of the days of the week nevertheless preserve the seven-day rhythm and stick to the traditional practice of observing the Sabbath every seventh day, despite the likelihood of its being the wrong day in historical time.[8])

The week is all around us. Yet, to this day, it has never been the single focus of a comprehensive scholarly investigation. Paradoxically, it has managed to escape our attention for so long precisely because it is ubiquitous. After all, it is very often the most familiar and "obvious" aspects of our environment that are the least studied ones. "Nothing evades our attention so persistently as that which is taken for granted. As a rule, we notice explicitly only those features of our total experience which strike our attention by their *not* being obvious. . . . Obvious facts tend to remain invisible."[9] The seven-day week is definitely one such fact. The main goal as well as the greatest challenge of the present book is, therefore, to focus on one of the most familiar cornerstones of everyday life, which is usually overlooked as "obvious," and make it more explicitly "visible."

The book traces the origins of the seven-day week in both Judaism and astrology and examines the social history of its diffusion throughout the world. Of particular concern are the way in which the three major monotheistic religions (Judaism, Christianity, and Islam) have managed to distance themselves from—and accentuate their distinctiveness vis-à-vis—one another through the establishment of three separate weekly cycles "peaking" on Saturday, Sunday, and Friday, respectively, as well as the two most serious, if unsuccessful, attempts in history to destroy the seven-day "beat" of human life. The book then examines various attempts throughout history to synchronize the week with both the month and the year and fix permanently the relations between particular dates and particular days of the week. Next it explores the way in which, through imposing a rhythmic "beat" on a vast array of major activities (including work, consumption, and socializing), the week promotes the structuredness and orderliness of human life, making it more regular and thus more predictable. Finally the book highlights the way in which a regular weekly schedule constitutes an ideal framework for organizing the compartmentalization of modern life.

Equating the week with a seven-day rhythm is a result of an ethnocentric bias that is challenged through an examination of the surprisingly wide variability of the week's length in different parts of the world. (Such an examination also highlights the variability of the functions of the week, from the regulation of religious or economic activity to the construction of divinatory calendars.) This comparative perspective on the weekly "beat" helps unveil its conventional nature. The overwhelming variability in the rhythm of activity within a single species clearly cannot be attributed to nature. It is obviously a product of human interference with the natural order of things.

In helping to shed more light on the fundamental, yet rather murky, difference between natural inevitability and social conventionality, the week provides an ideal context for examining the distinctively human interaction with time. Despite its apparent inevitability, it is not a part of nature, but, rather, a cultural artifact that rests on social convention alone. Unlike the day and the year, the week is an artificial rhythm that was created by human beings totally independently of any natural periodicity.

While nature has given us the day and the month, it has not been of much help in our attempts to establish regular patterns that do not repeat so frequently as "every day" yet not so infrequently as "every month." Various civilizations, independently of one another and often for entirely different purposes, have managed to fill this natural gap by inventing different variants of such a cycle. The week is the only major rhythm of human activity that is totally oblivious to nature, resting on mathematical regularity alone. Its invention was one of the first major attempts by humans to break away from being prisoners of nature and create an artificial world of their own, and therefore ought to be regarded as one of the greatest breakthroughs in the history of human civilization.

CHAPTER ONE

The Origins of the Seven-Day Week

For most of us, "week" is synonymous with "seven days." In fact, in many languages, the word "week" is either identical to, or directly derives from, the word "seven":

shavu'a	Hebrew
isbu'u	Arabic
hebdomas	Greek
haftah	Persian
yo't'neag	Armenian
sedmica	Serbo-Croatian
hét	Hungarian
sizun	Breton
seithan	Cornish
seachduin	Gaelic

That is also true of the Latin word for "week," *septimana*, from which the following words have derived:

săptămână	Rumanian
settimana	Italian
setmana	Catalan
semana	Spanish, Portuguese
semaine	French

The question "why seven?" seems almost inevitable. Since the seven-day week is not a natural cycle, the only way to explain its length would be by considering its origins and the social history of the diffusion of this cycle throughout the world.

From an historical standpoint, there are two ways of explaining why the week to which we adhere is seven days long, neither of which necessarily excludes the other. One explanation relates the length of our week to the seven days of the Creation in traditional Jewish cosmology, while the other relates it to the seven planets of ancient astrology. Quite coincidentally, both Judaism and astrology originated in western Asia, and it is from that region—largely through the help of Christianity, Hinduism, and Islam—that the seven-day week has spread into Europe, Africa, the rest of Asia, the Americas, and Oceania.[1] If there are any adult human beings today who are not familiar with this cycle, they must inhabit one of those increasingly fewer "remote" parts of the world where Judaism, Christianity, Hinduism, and Islam have not yet made their overwhelming impact.

The Jewish Week

The Hebrew word for "week," *shavu'a*, dates from the time of the Old Testament,[2] yet, back in antiquity, it was often used quite interchangeably with the word for the Sabbath, *shabbath*. Thus, for example, while one biblical passage instructs the ancient Israelites to celebrate the Feast of First Fruits (literally referred to there as the Feast of Weeks) seven weeks after Passover, a parallel passage suggests that seven sabbaths ought to be counted instead.[3] Centuries later, in the Talmudic and Midrashic literature, the week was still referred to as *shabbath*.[4] It is quite clear then that the periodic observance of the Sabbath every seven days was at the heart of the Jewish week from the very beginning.

For those who take the biblical account of the Creation both seriously and literally, the length of the seven-day week presents no problem at all. The practice of working for six days and then resting periodically on the seventh, which appears to be the main raison d'être for the institutionalization of this cycle, is essentially believed to have originally been a divine temporal pattern which requires no further explanation. It was first practiced by God when creating the universe: "And on the seventh day God finished His work which He had made; and He rested on the seventh day from all His work which He had made. And God blessed the seventh day, and hallowed it; because that in it He rested from all His work which God in creating had made."[5] The eternal imitation of God's rest on "the seventh day" following the six days of the Creation is the main theological account of the Jewish weekly Sabbath rest, from which the seven-day week has evolved. That claim is made quite explicitly in the Ten Commandments: "Remember the sabbath day, to keep it holy. Six days shalt thou labor, and do all thy work; but the seventh day is a sabbath unto the Lord thy God. . . . for in six days the Lord made heaven and earth, the sea, and all that in them is, and rested on the seventh day; wherefore the Lord blessed the sabbath day, and hallowed it."[6] The connection is later made even more clearly:

> Six days shall work be done; but on the seventh day is a sabbath of solemn rest, holy to the Lord; whosoever doeth any work in the sabbath day, he shall surely be put to death. Wherefore the children of Israel shall keep the Sabbath, to observe the sabbath throughout their generations, for a perpetual covenant. It is a sign between Me and the children of Israel for ever; for in six days the Lord made heaven and earth, and on the seventh day He ceased from work and rested.[7]

(This in itself, incidentally, still does not explain the evolution of a continuous seven-day cycle. It has been argued, for example, that the Sabbath was originally the seventh day of the year and was observed, upon the conclusion of a six-day commemoration of the Creation, only once a year.[8])

The word *shabbath* may have originally derived from *sh-b-th*, a verb which literally means "to cease from labor" and which, in the above passage from Genesis, is used to refer to God's rest following the Creation. In fact, the way this verb is used in some other biblical passages seems to suggest that the Sabbath observance and the periodical abstention from work may have been originally inseparable from one another conceptually.[9] The word *shavu'a*, however, is etymologically as well as conceptually related to the Hebrew word for "seven," *sheva*. In fact, throughout the *Book of Jubilees*, it is used to refer to seven-year, rather than seven-day, periods. As the concept "sabbatical year" seems to suggest, this connection may have originally also been made with regard to *shabbath*.

Seven-day intervals played a prominent role in ancient Jewish liturgy and ritual,[10] yet numerous passages throughout the Old Testament[11] seem to indicate a much broader symbolic significance of the number 7. That, however, was also true of the ancient civilizations of Mesopotamia, where the number seven played a prominent role in liturgy, ritual, magic, and art. The ancient Babylonians, for example, regarded the universe as a sevenfold entity governed by a fusion of seven deities. Consequently, seven-day intervals came to symbolically represent the ideas of totality and completeness, and were generally regarded as homogeneous, closed periods of time.[12] These intervals were possibly incorporated into the ancient West Asian calendrical system, either through the dedication of each day to one of seven winds, or as the building blocks of the "pentecontad" calendar, an agricultural calendar which has been preserved in part in both Judaism and Christianity and which survived in southern Palestine as late as the beginning of this century.[13] This calendar was essentially based on fifty-day intervals, which were possibly divided into seven seven-day intervals plus an additional day known as *atzeret*. It was probably the elimination of the *atzeret* days from the calendar that eventually led to the evolution of a continuous seven-day cycle.[14]

The designation of the seventh, fourteenth, nineteenth, twenty-first, and twenty-eighth days of a lunar month in a religious Assyrian calendar from the seventh century B.C. as "evil days" (*umu limnu* or *uhulgallu*) provides some further evidence of a possible non-Jewish origin of the Sabbath observance.[15] The prohibitive character of days that are spaced seven days apart from one another, including such precautionary measures as abstention from chariot riding and eating cooked meat on the part of the king, bears close resemblance to that of the Sabbath, particularly given the strict traditional taboo on traveling and cooking on that day.

The non-Jewish origin of the Sabbath seems even more plausible upon the realization that there is actually no conclusive historical evidence that Jews had indeed observed the Sabbath regularly every seven days prior to the Exile, when they first came into close contact with the dwellers of Mesopotamia![16] Whereas all the major Jewish annual pilgrimage festivals had originally been associated with the Temple in Jerusalem, the Sabbath is most probably a product of the Exile period in Jewish history, which followed the destruction of the Temple by the Babylonians in 586 B.C. Incidentally, this period is also known as the "Synagogue period," after the institution that was established in exile to partly replace the lost Temple, and it is important to note that the evolution of the synagogue is historically associated with that of the Sabbath. As late as the first century B.C., the synagogue was still known as "the Sabbath house,"[17] which, unlike the Temple, was deliberately designed from the very beginning for weekly congregation.

Despite all of the above, the establishment of a seven-day week based on the regular observance of the Sabbath is a distinctively Jewish contribution to civilization. The choice of the number 7 as the basis for the Jewish week might have had an Assyrian or Babylonian origin, yet it is crucial to remember that the ancient dwellers of Mesopotamia themselves did not have a real seven-day week.[18] While it is possible that the seven-day intervals entailed in the regular observance of the seventh, fourteenth, twenty-first, and twenty-eighth days of the lunar month might have served as the model for the Jewish week, they themselves cannot be considered weeks. Such intervals, which I shall call *quasi weeks*, undoubtedly bear a striking resemblance to the week and are often mistaken for it. Nevertheless, they are an essentially different phenomenon.

One of the most distinctive features of the week is the fact that it is entirely dissociated from the lunar cycle. It is essentially defined as a precise multiple of the day, quite independently of the lunar month. Quasi weeks, on the other hand, are generally defined as rough approximations of fractions of the lunar month, and are appropriately called "lunar weeks" by Francis H. Colson.[19] From a calendrical standpoint, there is a considerable affinity between the Assyrian observance of the "evil days" and the ancient Persian observance of the first, eighth, fifteenth, and twenty-third days of the lunar cycle, or the Buddhist observance of the *uposatha* on the new and full moons as well as on the days that are eight days away from both. Consider also, in this regard, the practice of subdividing the lunar month into quasi-weekly cycles other than seven days long—three 10-day intervals (in ancient China and Greece as well as among the Ahanta of Ghana and the Maori of New Zealand), four 8-day ones (in northern Ethiopia), six 5-day ones (among the Wachagga of Tanzania), and so on.[20]

Those who believe that our seven-day week has derived from the lunar cycle seem to forget that the latter is not really a twenty-eight-day cycle. In fact, approximately twenty-nine days, twelve hours, forty-four minutes, and three seconds—that is, about 29.5306 days—elapse between any two successive new moons. (That should also preclude any lunar origin of the fortnight, which literally means "fourteen nights." One half of the lunar cycle is actually much closer to fifteen than to fourteen days.) The lunar month clearly cannot be divided in a "neat" manner into weekly blocks of complete days. Any subdivision of the lunar cycle necessarily involves some mathematically inconvenient remainder of hours, minutes, and seconds. A precise quarter of the lunar cycle, for example, amounts to 7.38625 days, and any week of that length would necessarily have to begin at different times of the day.

Hence the essential irregularity of quasi-weekly rhythms. In order to always begin at the same time of day, quasi weeks obviously cannot be of a standard length. For example, in order to fit into the standard north Ethiopian thirty-day lunar calendar month, the last two days of every fourth eight-day *'sāmĕn* inevitably had to be eliminated.[21] Along similar lines, given that some lunar calendar months are only twenty-nine, rather than thirty, days long, many of the ancient Chinese *hsüns* and Greek *decades*, which were normally ten days long, had to be only nine days long. Also note, in this regard, that, despite their striking superficial calendrical resemblance to the Sabbath, the ancient Assyrian "evil days" and the Buddhist *uposatha* were not always spaced seven days apart from one another, since four seven-day cycles would have amounted to only twenty-eight, rather than 29.5306, days.

In short, being tied to the lunar cycle, quasi weeks are necessarily irregular, since the completion of every lunar month essentially interrupts their continuous flow. To appreciate the discontinuity inherent to living in accordance with quasi-weekly rhythms, consider what it would be like if the continuous flow of our own week were to be interrupted every month or so by the elimination of a day or two. Imagine, for example, having to cancel regular Sunday family gatherings or Tuesday classes every month or so due to the lack of a Sunday or a Tuesday on that particular week!

Such a state of affairs might not be considered problematic by the Bontoc Igorot of the Philippines, who used to observe their sacred rest day "on the average, about every ten days during the year, though not with absolute regularity."[22] The discontinuity inherent to quasi weeks is very disruptive, however, to any society concerned with *temporal regularity*. Accomplishing such regularity essentially involves rigidifying the rate of recurrence of periodic activities,[23] which, in turn, presupposes a uniform duration of the various cycles along which human life is temporally structured.

Hence the indispensability of a *continuous* week for the establishment of settled life with a high level of social organization, particularly significant since the rise of a market economy, which involved orderly contact on regularly recurrent, periodic market days.[24] Only by establishing a weekly cycle of an unvarying, standard length could society guarantee that the continuity of its life would never be interrupted by natural phenomena such as the lunar cycle. The dissociation of the week from the lunar cycle, is, therefore, the most significant breakthrough in the evolution of this cycle from its somewhat rudimentary and imperfect predecessor. Only by defining the week as a precise multiple of the day, rather than as a rough approximation of a fraction of the lunar month, could human beings permanently avoid the problem of having to handle loose remainders and, thus, introduce into their lives the sort of temporal regularity that they could never attain with the quasi week.

The first people to have established a continuous weekly cycle that was entirely independent of the lunar cycle were the ancient

Egyptians.[25] Possibly as a result of being sun-worshipers, which essentially freed them from the necessity of observing lunar rites, they practically ignored the moon in their civil calendar, which was based on twelve 30-day months, each of which was, in turn, subdivided into three 10-day weeks. (The additional five "epagomenal" days would thus interrupt the continuous flow of the week only once a year.) Through marking the beginning of each week by the rising of the main star of a particular celestial constellation, they managed to establish a perfect harmony between the thirty-six constellations of the heavens and the thirty-six weeks of the civil calendar. And yet, having no clear evidence that this association between the temporal subdivisions of the calendar year and the spatial subdivisions of the heavens had any significance other than astrological, we must return to ancient Judaism for an early instance of a continuous, nonlunar week used for establishing and regulating actual rhythms of human activity.

It is interesting to note that the rise of the Sabbath cult within Judaism coincided with the withdrawal from worshiping the celestial bodies, and particularly the moon.[26] In other words, the dissociation of the week from a natural cycle such as the waxing and waning of the moon can be seen as part of a general movement toward introducing a supranatural deity. Not being personified as any particular natural force, the Jewish god was to be regarded as untouched by nature in any way. Accordingly, the day dedicated to this god[27] was to be regarded as part of a divine temporal pattern that transcends even nature itself. That obviously involved dissociating the week from nature and its rhythms. Only by being based on an entirely *artificial mathematical rhythm* could the Sabbath observance become totally independent of the lunar or any other natural cycle.

A continuous seven-day cycle that runs throughout history paying no attention whatsoever to the moon and its phases is a distinctively Jewish invention.[28] Moreover, the dissociation of the seven-day week from nature has been one of the most significant contributions of Judaism to civilization. Like the invention of the mechanical clock some 1,500 years later, it facilitated the establishment of what Lewis Mumford identified as "mechanical periodicity,"[29] thus essentially increasing the distance between human beings and nature. Quasi weeks and weeks actually represent two fundamentally distinct modes of temporal organization of human life, the former involving partial adaptation to nature, and the latter stressing total emancipation from it. The invention of the continuous week was therefore one of the most significant breakthroughs in human beings' attempts to break away from being prisoners of nature and create a social world of their own.[30]

The Astrological Week

The origins of our week are far more complex, however. A brief glance at the names of the days of the week in English as well as various other European languages serves to remind us that the seven-day cycle was originally associated not only with the seven days of the Creation, but also with the seven so-called "planets." All those listed in Figure 1 have derived from the original Latin names of the days of the week, the literal translations of which are "the day of Saturn" (Saturday), "the day of the sun" (Sunday), "the day of the moon" (Monday), "the day of Mars" (Tuesday), "the day of Mercury" (Wednesday), "the day of Jupiter" (Thursday), and "the day of Venus" (Friday). In the case of most of the Celtic and Romance languages, these derivations have been simply phonetic. In the case of most of the Germanic languages, however, there was an actual process of "translation," which involved some association between the original Roman gods and goddesses and their Nordic "counterparts." The day of Mars could thus become the day of the war god Tyr, "the day of Mercury" was translated into "the day of Woden-Odin,"[31] "the day of Jupiter" into "the day of Thor-Donar-Thunar," and "the day of Venus" into "the day of Fria-Frigg."

To the ancient stargazers, the most striking feature of Saturn, the sun, the moon, Mars, Mercury, Jupiter, and Venus was the fact that they seemed to be moving regularly across the heavens. (While we now consider their assumption regarding the movement of the sun as erroneous, we should nevertheless remember that, from their own geocentric perspective, the sun indeed did move. From that perspective, there was also no significant difference between the movement of the moon and that of Mars or Saturn.) In order to set them apart from all the other celestial bodies, which appeared to be stationary, the ancients grouped those seven luminaries together, and appropriately called them "planets," a derivative from the Greek verb *planasthai*, which literally means "to wander."

The original clustering of the seven "wanderers" together was the work of the Babylonians—also known as Chaldeans—who were the first to have observed the heavens with the systematic rigor necessary for the development of astronomy.[32] However, it is not as the pioneers of the science of astronomy that the Chaldeans enter our story, but, rather, as those who laid the foundations of the "occult" system of astrology. (The words "astrologer" and "Chaldean" were synonymous in the classical world, and, to this day, the latter is still used to refer to a person versed in the occult arts.) In fact, it is quite possible that those ancient dwellers of Mesopotamia actually developed their astronomy—including numerous tables for the metic-

FIGURE 1 Days of the Planets

LANGUAGE	SATURN	SUN	MOON	MARS	MERCURY	JUPITER	VENUS
Latin	dies Saturni*	dies Solis	dies Lunae	dies Martis	dies Mercurii	dies Jovis	dies Veneris
Cornish	de Sadarn	de Sil	de Lûn	de Merh	de Marhar	dê Jeu	de Gwenar
Breton	Disadorn	Disul	Dilun	Dimeurz	Dimerc'her	Diriaou	Digwener
Welsh	dydd Sadwrn	dydd Sul	dydd Llun	dydd Mawrth	dydd Mercher	dydd Iau	dydd Gwener
Gaelic	Di-sathuirne		Di-luain	di Màirt			
Catalan			Dilluns	Dimarts	Dimecres	Dijous	Divendres
French			Lundi	Mardi	Mercredi	Jeudi	Vendredi
Italian			Lunedi	Martedi	Mercoledi	Giovedi	Venerdi
Spanish			Lunes	Martes	Miércoles	Jueves	Viernes
Rumanian			Lunì	Martì	Miercurì	Joì	Vinerì
English	Saturday	Sunday	Monday	Tuesday	Wednesday	Thursday	Friday
Swedish		Söndag	Måndag	Tisdag	Onsdag	Torsdag	Fredag
Danish		Søndag	Mandag	Tirsdag	Onsdag	Torsdag	Fredag
Norwegian		Søndag	Mandag	Tirsdag	Onsdag	Torsdag	Fredag
Icelandic		Sunnundagur	Mánudagur				
Finnish		Sunnuntai	Maanantai	Tiistai		Torstai	
Lapp			Manodag	Tisdag		Tuoresdag	
Dutch	Zaterdag	Zondag	Maandag	Dinsdag	Woensdag	Donderdag	Vrijdag
German		Sonntag	Montag	Dienstag		Donnerstag	Freitag
Hungarian		Vasárnap					
Albanian	e Shtunë	e Dielë	e Hënë	e Martë	e Mërkurë	e Enjte	

* Foreign proper names appear in roman type and are capitalized according to the conventions of English, throughout this book.

ulous calculation of planetary positions—primarily for the purpose of designing horoscopes. The main Chaldean contribution to the evolution of the astrological seven-day week was the development of the so-called "planetary theory," which involves a belief in the worldly influence of the seven "wanderers." That each of those seven planets (generally conceived as deities) affects human fortune in its own distinctive way is originally a Babylonian idea.

However, while the Chaldean origin of astrology is indisputable, there is no evidence that an actual astrological seven-day cycle ever existed in ancient Mesopotamia.[33] While the planetary theory most probably evolved around 500 B.C. and the earliest Babylonian horoscope dates from 409 B.C.,[34] not one instance of a particular day being designated as "the day of the moon" or "the day of Venus," for example, has yet been found in pre-Hellenistic horoscopes.

While its origins—just like those of the Jewish seven-day week—may very well lie in Mesopotamia, the astrological seven-day week actually came into being only in the aftermath of Alexander the Great's conquest of western Asia, and was essentially a Hellenistic invention. It most probably evolved sometime during the second century B.C. at the very heart of the Hellenistic world, namely Alexandria.[35] (This might explain why our main classical source regarding this cycle attributed its invention to the Eygptians.[36]) It was only in Alexandria that three distinct practices that had evolved quite independently of the Chaldean planetary theory—an astronomical practice of arranging the seven planets in a certain invariable order, a mathematical practice of subdividing the daily cycle into twenty-four hours, and an astrological theory known as the doctrine of "chronocratories"—were nevertheless integrated with it so as to produce the astrological seven-day week in its final form. This cycle is therefore the product of the successful Hellenistic fusion of astronomy, astrology, and mathematics, as well as of the great cultural heritage of Egypt, Babylonia, and Greece.

The sequence of the seven days of the astrological week is essentially based on the arrangement of the seven planets in the fixed, invariable order Saturn–Jupiter–Mars–Sun–Venus–Mercury–Moon, a distinctively Hellenistic arrangement that evolved only in the second century B.C.[37] This planetary sequence corresponds to the order of the orbital periods of the planets—Saturn's orbital period is approximately 29.46 years, Jupiter's 11.86 years, Mars's 686.98 days, the sun's (given the assumption that it is the sun that revolves around the earth rather than the other way around) 365.26 days, Venus's 224.70 days, Mercury's 87.97 days, and the moon's 29.53 days.[38] It may have also derived, however, from the astronomical theory of "geocentric distances," according to which planets were arranged in the order of their distance from earth, with Saturn and the moon being the farthest and nearest planets, respectively.[39] (Modern astronomical knowledge, of course, would have required the reversal of the order of Venus and Mercury.)

That still does not explain why "the day of the moon," for example, always follows "the day of the sun" rather than "the day of Mercury." If we examine how the sequence of the seven days of the astrological week might have derived from the order in which the Hellenistic astronomers arranged the seven planets, we are immediately struck by one obvious overall pattern. In order to derive the sequence of the planetary days from the series Saturn–Jupiter–Mars–Sun–Venus–Mercury–Moon, one must proceed by "planetary leaps" that consist of skipping two planets each time. Thus, beginning with Saturn, we would proceed to the sun (through skipping Jupiter and Mars), from there to the moon (through skipping Venus and Mercury), from there to Mars (through skipping Saturn and Jupiter), and so on. If we assigned days to each of the planets where we stopped, we would arrive at the following series: the day of Saturn, the day of the sun, the day of the moon, the day of Mars, the day of Mercury, the day of Jupiter, and the day of Venus. As we have already seen, that is precisely the order of the days of the astrological week.

Whence comes this "planetary leap"? Unfortunately, a piece written around A.D. 100 by the Greek essayist Plutarch and bearing the suggestive title "Why Are the Days Named After the Planets Reckoned in a Different Order from the Actual Order?"[40] has been lost forever, and our earliest source on this matter is the third-century Roman historian Dio Cassius.[41] One explanation he claims to have heard relates to the tetrachord, a diatonic series of four tones with an interval of a perfect fourth between the first and last tones. The principle underlying the process of constructing intervals of perfect fourths happens to be identical to the one we have just discovered in the case of the days of the astrological week, namely, proceeding through the cycle of seven tones by skipping two tones each time. Given the interdependence of astronomy, mathematics, and music—not to mention the mystical significance of the number seven—in Pythagoreanism, I would suspect that this explanation probably originated within that ancient mystical system.[42] The problem with this explanation, however, is that it ignores the entire astrological context within which the planetary seven-day week evolved.[43] It is necessary, therefore, to look at the other major Hellenistic contribution to the evolution of this cycle, namely, the astrological doctrine of "chronocratories," which presupposes the subdivision of the daily cycle into twenty-four "planetary" hours.

The length of the astrological week was largely a result of the fact that the ancient Babylonian astronomers happened to identify seven planets. (Had they been able, with the help of some sophisticated telescopes, to also observe Uranus, Neptune, and Pluto, the week might have evolved as a ten-day cycle. On the other hand, had they accepted our heliocentric model of the planetary system, where the sun is considered a stationary star rather than a planet, this cycle might have been only six days long.) Yet the length of this astrological cycle was also affected by the interaction between those seven luminaries and the twenty-four hours of the day. The product of that interaction was a cycle that was not only seven days, but also 168 hours, long. It was only once every 168 hours—168 being the lowest common product of both 7 and 24—that the daily 24-hour cycle and a 7-hour cycle that we shall soon encounter would be completed together.

In order to fully appreciate the significance of the subdivision of the daily cycle into twenty-four hours—a distinctively Egyptian mathematical practice—for the evolution of the astrological week, it is important to realize that, from an astrological standpoint, the hour is at least as essential a unit of time as the day.[44] (Consider, in this respect, the etymology of the word "*horo*scope.") In fact, the ancient astrologers came to designate particular days as "the day of Mars" or "the day of the moon" only after first having designated particular hours as "belonging" to those planets. According to the astrological doctrine of "chronocratories," which most probably also originated in Alexandria, each one of the twenty-four hours of the daily cycle was "assigned" to one of the seven planet-deities, which was regarded as its "chronocrator" or "controller." In addition, the controller of the first hour of each day was also designated as the "regent" of that entire day as a whole.[45] Each moment of a person's life was thus believed to be astrologically dominated by both the controller of the hour and the regent of the day within which it was anchored.

Dio Cassius claimed to have also heard a second explanation of the order of the planetary days. This explanation is much more convincing, since it takes into account not only the traditional Hellenistic arrangement of the planets in a sequence corresponding to their orbital periods or assumed geocentric distances, but also the Hellenistic doctrine of chronocratories, which provides the astrological context within which the planetary seven-day week actually evolved. According to this explanation, one would astrologically "assign" the first hour of the first day to the most distant planet, Saturn, and then proceed to assign each following hour to the following planet in the traditional sequence Saturn–Jupiter–Mars–Sun–Venus–Mercury–Moon. The second hour of the first day would thus be assigned to Jupiter, the third one to Mars, and so on (see Figure 2). As there were only seven planets, the eighth hour would again be assigned to Saturn, and the seven-hour cycle would begin anew. The twenty-fifth hour of the first day would have been assigned to the sun, yet the daily cycle was divided into only twenty-four hours, so it would be the first hour of the second day. Since the controller of the first hour of each day was also supposed to dominate that entire day as a whole, the entire first day came to be astrologically assigned to Saturn, the second one to the sun, the third to the moon, the fourth to Mars, the fifth to Mercury, the sixth to Jupiter, and the seventh to Venus. At that point the 168-hour cycle would be completed and the regent of what would have otherwise been the eighth day would once again be Saturn, the regent of the first day of the cycle.

Hence the above-mentioned "planetary leap," whereby, in order to arrive at the sequence of planetary days, one would have to proceed along the traditional planetary sequence by skipping two planets and stopping only at the third one. Since there were seven planets, each 24-hour daily cycle would include three additional hours beyond the three complete series of seven "planetary" hours. In other words, the "planetary leap" from the regent of any one particular day to that of the following one was three planets long because three is essentially what one would get by subtracting three complete seven-hour cycles from the twenty-four-hour daily cycle.

The Jewish and astrological weeks evolved quite independently of one another. However, given the coincidence of their identical length, it was only a matter of time before some permanent correspondence between particular Jewish days and particular planetary days would be made. A permanent correspondence between the Sabbath and "the day of Saturn" was thus established no later than the first century of the present era,[46] and Jews even came to name the planet Saturn Shabtai, after the original Hebrew name of the Sabbath, Shabbath. Moreover, as they came into closer contact with Hellenism, their conception of their holy day was evidently affected by the astrological conception of Saturn as a planet that has an overwhelming negative influence[47] (a conception which, incidentally, is still evident even from the association of the English word "saturnine" with a gloomy disposition). There are traditional Jewish superstitious beliefs about demons and evil spirits that hold full sway on the Sabbath, and an old Jewish legend even links the choice of "the day of Saturn" as the official Jewish rest day with the superstition that it would be an inauspicious day for doing any work anyway![48]

And yet, even if Judaism was indeed influenced by the astrological lore regarding the unlucky character of Saturday, the Sabbath

FIGURE 2 **The Astrological Seven-Day Week**

1 SATURN	20 Jupiter	14 Jupiter	8 Jupiter
2 Jupiter	21 Mars	15 Mars	9 Mars
3 Mars	22 Sun	16 Sun	10 Sun
4 Sun	23 Venus	17 Venus	11 Venus
5 Venus	24 Mercury	18 Mercury	12 Mercury
6 Mercury		19 Moon	13 Moon
7 Moon	1 MOON	20 Saturn	14 Saturn
8 Saturn	2 Saturn	21 Jupiter	15 Jupiter
9 Jupiter	3 Jupiter	22 Mars	16 Mars
10 Mars	4 Mars	23 Sun	17 Sun
11 Sun	5 Sun	24 Venus	18 Venus
12 Venus	6 Venus		19 Mercury
13 Mercury	7 Mercury	1 MERCURY	20 Moon
14 Moon	8 Moon	2 Moon	21 Saturn
15 Saturn	9 Saturn	3 Saturn	22 Jupiter
16 Jupiter	10 Jupiter	4 Jupiter	23 Mars
17 Mars	11 Mars	5 Mars	24 Sun
18 Sun	12 Sun	6 Sun	
19 Venus	13 Venus	7 Venus	1 VENUS
20 Mercury	14 Mercury	8 Mercury	2 Mercury
21 Moon	15 Moon	9 Moon	3 Moon
22 Saturn	16 Saturn	10 Saturn	4 Saturn
23 Jupiter	17 Jupiter	11 Jupiter	5 Jupiter
24 Mars	18 Mars	12 Mars	6 Mars
	19 Sun	13 Sun	7 Sun
1 SUN	20 Venus	14 Venus	8 Venus
2 Venus	21 Mercury	15 Mercury	9 Mercury
3 Mercury	22 Moon	16 Moon	10 Moon
4 Moon	23 Saturn	17 Saturn	11 Saturn
5 Saturn	24 Jupiter	18 Jupiter	12 Jupiter
6 Jupiter		19 Mars	13 Mars
7 Mars	1 MARS	20 Sun	14 Sun
8 Sun	2 Sun	21 Venus	15 Venus
9 Venus	3 Venus	22 Mercury	16 Mercury
10 Mercury	4 Mercury	23 Moon	17 Moon
11 Moon	5 Moon	24 Saturn	18 Saturn
12 Saturn	6 Saturn		19 Jupiter
13 Jupiter	7 Jupiter	1 JUPITER	20 Mars
14 Mars	8 Mars	2 Mars	21 Sun
15 Sun	9 Sun	3 Sun	22 Venus
16 Venus	10 Venus	4 Venus	23 Mercury
17 Mercury	11 Mercury	5 Mercury	24 Moon
18 Moon	12 Moon	6 Moon	
19 Saturn	13 Saturn	7 Saturn	1 SATURN

observance had been established long before the astrological week even came into being, and obviously preceded both the making of the calendrical association between "the day of Saturn" and the Sabbath and the naming of the planet after the latter.[49] Whereas astrology named the day of Saturn after the planet, Judaism did that the other way around.

By the early third century of the present era, when Dio Cassius wrote his *Roman History,* the astrological week had already "spread to all men" and was "becoming quite habitual to all the rest of mankind and to the Romans themselves."[50] A list of the seven planetary days in a grammar textbook dating from A.D. 207, and an inscription from A.D. 205, where the date is designated not only by the year and the month but also as "the day of the moon," attest the wide popularity of that cycle in Rome during that period.[51] Yet references to planetary days appear in the writings of church fathers already during the second century,[52] and a calendar dating from the time of Emperor Trajan, who died in A.D. 117, depicts the planetary deities in the unmistakable order of the planetary days.[53] Furthermore, a wall inscription listing the seven planetary deities under the heading "days" and a fresco depicting the figures representing them have both been unearthed in Pompeii,[54] and we know they could not have been made later than A.D. 79, the year that city was destroyed. Not only are the planetary deities listed there under the heading "days." In both instances they are also arranged in the sequence Saturn–Sun–Moon–Mars–Mercury–Jupiter–Venus, which—particularly given the customary arrangement of the planets in the Hellenistic sequence Saturn–Jupiter–Mars–Sun–Venus–Mercury–Moon during that period—would have made no sense whatsoever had the astrological week not been in use! Yet the latter appears to have been introduced to Rome even before the beginning of the present era. A cycle of seven days designated by the letters A through G appears in the *fasti Sabini,* fragments of a calendar dating from sometime between 19 B.C. and A.D. 4,[55] whereas an explicit reference to "the day of Saturn" appears in a verse written by the poet Tibullus, who died in 19 B.C.[56]

We can conclude that the astrological seven-day week, which evolved in Alexandria during the second century B.C., was introduced to the West through Rome sometime toward the end of the first century B.C. If it was Alexander the Great's conquest of Greece, Babylonia, and Egypt that, in bringing those three civilizations together, was indirectly responsible for the evolution of the astrological week in the first place, it was Julius Caesar's conquest of Egypt that, in making Rome heir to the glorious Hellenistic heritage, was responsible for importing that oriental cycle to the Occident.

The diffusion of the seven-day week throughout the Roman Empire was helped by the growing popularity of astrology in Rome in the aftermath of that historic meeting between East and West. Yet astrology was not the only movement to have come from the Orient and capture the Western imagination; soon after, Christianity arrived. As we shall now see, it was the Church that was responsible for integrating the Jewish and astrological weeks together and spreading the seven-day cycle throughout most of the world.

The Church and the Diffusion of the Week

The original significance of the week for Christianity is preserved to this day in the word for "Sunday" in several European languages:

Domenica	Italian
Duminică	Rumanian
Domingo	Spanish, Portuguese
Di-dòmhnuich	Gaelic
Dimanche	French
Diumenge	Catalan

All these words derive from the Latin dies Dominica, a direct translation of the original Greek Kyriakě (from which both the Armenian Giragi and the Romany Koóroki have also derived), which literally means "the Lord's Day."

The reason why Sunday is regarded by the Church as "the Lord's Day" may be found in its Russian name, Voskresen'e, which literally means "Resurrection." A classic example of a "commemoration of origins,"[57] the Christian observance of Sunday evokes one of the most significant moments in the origin myth of the Church, namely Christ's manifestation of his presence to his followers through rising from the dead and ascending to heaven. All four Gospels agree that this event took place on "the first day of the [Jewish] week,"[58] and the association between the Sunday observance and the Resurrection was repeatedly emphasized by the early church fathers. Thus, for example, The Epistle of Barnabas, a first-century apocryphal work attributed to the apostle who founded the church of Antioch, notes that Sunday is observed by the Church because, on that day, Christ rose from the dead and ascended to heaven.[59] Similarly, around the middle of the second century, Justin Martyr points out that Christians assembled on Sunday because, "the day before that of Saturn, He was crucified; and on the day after it, which is Sunday, He appeared to His apostles and disciples."[60] The Church may have originally celebrated the Holy Eucharist only once a year, on Easter,[61] yet it soon came to associate the commemoration of the Resurrection with a weekly rhythm and, more specifically, with Sunday. On "the first day of the week," we read in the New Testament, the disciples would assemble for the "breaking of the bread" and Paul would collect alms.[62]

Being Jewish, the early Christians knew Saturday as Shabbath,[63] and this original Hebrew word still resonates today in the word for "Saturday" in many languages that are spoken primarily by Christians:

Sábbato	Greek
Shapat'	Armenian
Subbota	Russian
Subota	Serbo-Croatian
Sobota	Polish, Czech, Slovak, Slovene
Sabato	Italian
Sabado	Portuguese
Sábado	Spanish
Dissabte	Catalan
Samedi	French
Sâmbăta	Rumanian
Szombat	Hungarian

(For languages spoken primarily by non-Christians, consider also the Persian Shambah and the Arabic Yom-es-sabt.) The New Testament also refers to Sunday by its original Hebrew name, "the first day of the week," and to Friday by the Greek word Paraskeuĕ, which was used during the first and second centuries by Jews dwelling in Greek-speaking communities, and which literally means "the day of Preparation" (obviously for the Sabbath).[64] The rest of the days of the week were originally also named by the Church in accordance with their temporal distance from the preceding Sabbath, following the Hebrew practice.[65] Monday was thus designated as "the second day after the Sabbath" (Secunda Sabbati, in Latin), Tuesday as "the third day after the Sabbath" (Tertia Sabbati), and so on. (This archaic nomenclature is still preserved, at least in part, in Armenian, Greek, Portuguese, and Icelandic.)

The early Christians were Jewish and their observance of Sunday originally did not conflict with their adherence to the Jewish week, which essentially revolves around the observance of Saturday. The observance of the Lord's Day originated not as a substitute for the Sabbath observance, but, rather, as an addition to it, and the early Christians used to observe both Sunday (as Christians) and Saturday (as Jews). (It might be added that, around the time Christianity was born, the practice of resting on the Sabbath was also imitated by many non-Jews as well.[66])

However, there came a point in the history of the Church when the identity of being both a Christian and a Jew was becoming increasingly difficult to maintain. Christians began to feel the need to establish their own exclusive identity as Christians, and that was accomplished partly through stressing their own distinctiveness vis-à-vis the group out of which they had originally sprung. It was at that point that the Sunday observance no longer seemed to suffice by itself, and the abandonment of the Sabbath observance was becoming a social necessity.

One of the most effective ways to accentuate social contrasts is to establish a *calendrical contrast.* Schedules and calendars are intimately linked to group formation,[67] and a temporal pattern that is unique to a group often contributes to the establishment of social boundaries that distinguish as well as actually separate group members from "outsiders." To take the example of distinctive weekly patterns, one of the most conspicuous cultural peculiarities of English Catholics that serves to emphasize their distinctiveness vis-à-vis non-Catholics is their Friday abstinence from meat.[68] Likewise, ironically, it is the Sabbath observance that has always served to express the uniqueness of Jews as a group, practically segregating them from, and preventing their actual assimilation into, their Gentile environment.[69] Given all this, the Church's decision to use its Sunday observance as a substitute for, rather than as a mere addition to, the Sabbath observance ought to be regarded—along with its later decision to calendrically segregate Easter from Passover[70]—as one of the most significant political moves made by the early Christians as a self-conscious group.

As the social and cultural distance between Judaism and Christianity was increasing, Christians' urge to establish a weekly cycle that would be considerably different from its Jewish counterpart was becoming stronger. In order to accentuate their distinctiveness vis-à-vis Jews (particularly as they could be easily mistaken for any of the many other Jewish sects that flourished around that period), they looked for "something distinctively their own,"[71] and thus chose to congregate regularly on a day other than the one on which Jews used to assemble. In other words, their choice of Sunday as a day other than Saturday was at least as significant, from a sociological standpoint, as their association of that day with the Resurrection!

It was Paul who first condemned the Sabbath observance,[72] yet it was Saint Ignatius who, toward the end of the first century, pioneered the movement toward substituting the Sunday observance for the Sabbath observance.[73] As was the case later with the calendrical segregation of Easter from Passover,[74] it was the "modernist" church of Rome that led the campaign toward the abandonment of the Sabbath, while the more traditionalistic Eastern churches chose to preserve the original Jewish calendrical practices. Thus, while observing the Lord's Day, many Eastern churches—the Ethiopian Church being the most outstanding example—also kept recognizing the holiness of the Sabbath.[75] However, as in the case of Easter, it was the Church of Rome that came out as the eventual victor, and, in the mid-360s, the Church officially abandoned the Sabbath observance. As the Synod of Laodicea summed it up in its twenty-ninth canon: "Christians shall not Judaize and be idle on Saturday, but shall work on that day; but the Lord's day they shall especially honour, and, as being Christians, shall, if possible, do no work on that day. If, however, they are found Judaizing, they shall be shut out from Christ."[76]

Despite its obvious effort to provide the week with a distinctively Christian content, the Church nevertheless chose to preserve its Jewish seven-day rhythmic form. This should not be taken too lightly, as it could have chosen to assemble regularly in accordance with the traditional Roman eight-day weekly cycle, which we shall soon encounter. After all, in abrogating the Sabbath, the Church also destroyed the raison d'être of the Jewish seven-day week.[77]

The preservation of the seven-day rhythm was, in part, the obvious result of the Church's deep, unconscious attachment to Judaism, as well as a pragmatic attempt to avoid alienating the rather significant Jewish component of its membership unnecessarily. However, a brief glance at the names of the days of the week in most European languages ought to remind us that the Jewish week is not the only context within which the evolution of the ecclesiastical seven-day cycle ought to be viewed. As we shall see, it was the convergence of both Jewish and astrological weeks around the time Christianity was being introduced into the Roman Empire that produced the seven-day cycle that has since spread throughout most of the civilized world.

While the Roman Catholic Church has officially clung to the traditional Jewish nomenclature of the days of the week, planetary designations of these days appeared as early as the second century in the writings of the church fathers,[78] and were popularly used by Christians at least since A.D. 269.[79] (The only significant organized Christian attempt to restore the original Hebrew nomenclature of the days of the week seems to have been the official elimination of the "heathen names" by the First General Assembly of Pennsylvania, obviously representing the spirit of the Society of Friends, between 1682 and 1706.[80] Incidentally, to this day, Quakers still call their Sunday schools "First-Day schools.") As we can tell from the etymologically curious fact that no planetary designations of days of the week are to be found in either Greek or any of the Slavic languages, it is only the Eastern Church that seems to have succeeded in suppressing the considerable influence of astrology.[81] Rome was obviously much less successful, as the planetary names of at least some of the days of the week in English, German, Dutch, Danish, Norwegian, Icelandic, Swedish, Finnish, Lapp, Hungarian, Albanian, Rumanian, Italian, French, Catalan, Spanish, Breton, Gaelic, Welsh, and Cornish seem to indicate.

It is clear from the etymological evidence that astrology had spread throughout the Roman Empire earlier, and probably much faster, than Christianity. Thus, by the early fourth century, when the Church finally gained control over the Empire, it was evidently too late for any serious ecclesiastical effort to fully eliminate the astrological associations of the seven days of the week.[82] (Note, in this regard, that, for the adherents of astrology, the planetary week was obviously much more than a mere time-reckoning system, as we can tell from their serious effort to "translate" Roman planetary deities into their Nordic equivalents.[83] Curiously enough, most probably as a result of its astrological association with either the chief Roman deity Jupiter or the Nordic thunder god Thor, Thursday used to be regarded by the Estonians, at least until the turn of this century, as holier than even Sunday![84]) As we can see in Figure 1, even at the very heart of the Roman Empire, where languages that have derived from Latin prevail, it is only with respect to the two "key" days of the Judeo-Christian week, namely Saturday (the Sabbath) and Sunday (the Lord's Day), that the Church has managed to supersede astrology. The astrological influence is obviously even more pronounced around the fringes of the Roman Empire, where Christianity arrived only much later. English, Dutch, Breton, Welsh, and Cornish, which are the only European languages to have preserved to this day the original planetary names of all the seven days of the week, are all spoken in areas that were free of any Christian influence during the first centuries of our era, when the astrological week was spreading throughout the Empire. Neither have these languages derived from either Greek or Latin, the languages most closely associated with the Church. That, incidentally, is also true of all the other languages that have preserved the planetary designations of at least one of the two "key" days of the Judeo-Christian week— German, Gaelic, Danish, Norwegian, Icelandic, Swedish, Finnish, Hungarian, and Albanian. (Note, in this respect, that, in the Scandinavian languages, while Saturday is not associated with Saturn, neither is it associated with the Sabbath.)

However, while the astrological week opens with "the day of Saturn," the Church has nevertheless managed to keep "the day of the sun" as the first day of the week. Yet that too may very well suggest the possible influence of some pagan veneration of the solar disk. After all, is it a mere coincidence that, around the turn of the third century, we find the Christian apologist Tertullian explicitly defending the Church against the accusation that its Sunday observance had actually originated in some pagan sun cult?[85]

Early references to Christ as the sun, the prevalence of his sunlike halo in Christian art, and the Church's decision to fix the commemoration of the Nativity on December 25 (a day traditionally celebrated by sun-worshipers as the annual "birth" of the sun following the winter solstice) all seem to point to some possible "solar" origin of Christianity. More specifically, particularly given the evangelical association of the Nativity with the Persian magi, they may indicate some possible early contact between Christianity and Mithraism, a Persian religion that flourished throughout the Roman Empire during the second and third centuries, and which actually revolved around a cult of the sun.[86] The correspondence between the Mithraist "day of the sun" and the day claimed by the Church to have been the day of the Resurrection may indeed have been purely coincidental. Nevertheless, its observance probably added to the Church's legitimacy among Mithraists, who observed it anyway.[87] Furthermore, particularly given the great popularity of Mithraism, it may have also contributed considerably to the Church's success in proselytizing pagans throughout the Roman Empire.

Incidentally, this "Persian connection" may also help to explain why the Hindu planetary week, in striking contrast to the Hellenistic original, has always begun on the day of the sun rather than on the day of Saturn.[88] Interestingly enough, the Indian days of the week (*vâras*) had already matched their European counterparts many centuries before regular contact between India and the West was established in the aftermath of the Age of Discovery, which is rather striking, considering how differently these two civilizations have always reckoned years, months, days of the month, and hours.[89] From ancient manuscripts we learn that the astrological week probably reached India sometime during the first century of our era.[90] From there, it also spread into Nepal, Tibet, Ceylon, Burma, Thailand, and Indochina.[91]

While India played the major role in the diffusion of the astrological week throughout South and Southeast Asia, it is the European colonization of Africa, the Americas, and Oceania that was responsible for introducing its Judeo-Christian cousin to large parts of these continents. Yet Christianity was by no means the only carrier that helped spread the Jewish week around the globe. Starting from the seventh century, Islam was responsible for importing this seven-day cycle to the east coast of Africa, the Sudan, Central Asia, large parts of North and West Africa, and even as far as to the Malay peninsula and parts of Indonesia.[92]

Mohammed's choice of Friday as the weekly day of Moslem public worship bears a striking resemblance to the Church's choice of Sunday as the Lord's Day. Like the early Christians, he also revealed his deep attachment to Judaism in a most conspicuous manner by adopting a seven-day religious rhythm for Islam. According to an ancient tradition, the early Moslems must have had the following argument: "The Jews have every seventh day a day, when they get together (for prayer), and so do the Christians; therefore, let us do the same."[93] And yet, at a time when Islam had to compete with the two other major monotheistic faiths for pagan proselytes, the Prophet probably also wanted to make sure that his followers would be properly distinguished, as well as actually segregated, from both Jews and Christians. Accordingly, he decided to differentiate the new Moslem weekly cycle from both Jewish and Christian weeks, choosing a day other than either Saturday or Sunday to be the principal weekly day of Moslem public worship: "Since the Christians had seized the day after the Saturday, he had no choice [!] but to take the day before it."[94] (It is not a mere coincidence, therefore, that Moslems have traditionally considered both Saturday and Sunday evil, or at least unfortunate, days.[95])

In his search for a day other than either Saturday or Sunday, Mohammed could have obviously settled on either Monday, Tuesday, Wednesday, or Thursday, rather than Friday. And yet, the day he chose as the "peak day" of the new religious week he established is—just like the Church's Lord's Day—a day that essentially touches the original weekly "peak day," namely the Sabbath. It is obviously not a mere coincidence that all three great monotheistic religions involve the use of weekly cycles of the very same length. Yet even more remarkable is the fact that these three cycles all happen to revolve around "peak days" that, while being different from one another, nevertheless all literally touch one another, just like the three religions they symbolically represent.

CHAPTER TWO

The Seven-Day Wars

T HE STORY of both Christian and Moslem weeks helps to shed some light on the political significance, as well as use, of the calendar, indicating how dramatic political changes are often accompanied by equally radical changes in the social structuring of time.[1] This "political" dimension of the weekly cycle will now be further explored.

While both Christianity and Islam have deliberately modified the internal structure of the original Jewish week by shifting its "peak," they have nevertheless both preserved its basic seven-day rhythmic form. However, throughout history, there have been a couple of very serious attempts to totally obliterate the seven-day "beat" through the introduction of alternative weekly cycles of an entirely different length altogether.

In the establishment of the length of the week and its diffusion throughout the world, religion was clearly the dominant force. The story of the two attempts to crush the seven-day week is, therefore, a story of Kulturkampf, the struggle of the modern state to overthrow traditional religious authority. Religion, however, may be the most resilient component of any tradition, and eliminating the seven-day week once it has been established as a regulator, as well as symbol, of religious life becomes next to impossible. The complete failure of the two boldest attempts to accomplish precisely that only makes the success of this seven-day religious cycle all the more impressive.

The French Ten-Day *Décade*

On December 20, 1792, the new assembly ruling France, the National Convention, authorized the Committee of Public Instruction to consider a general reform of the existing calendar. The committee essentially adopted a proposal originally made four years earlier by Pierre-Sylvain Maréchal,[2] and proceeded to recommend the establishment of a new calendar based on twelve new 30-day months, each of which would be divided into precisely three 10-day weekly cycles called *décades*.[3]

On October 10, 1793, only five days after the committee had presented its report to the convention, the official *Journal de France* discontinued its practice of dating issues using the traditional day of the week. Three weeks later, it began designating days by the committee's newly proposed names: Primidi, Duodi, Tridi, Quartidi, Quintidi, Sextidi, Septidi, Octidi, Nonidi, and Décadi. More significantly, on October 6, the convention had resolved to fix all public officials' rest days on Décadi. That resolution became an official decree on November 24, 1793, the day the entire new Revolutionary calendar was put into effect.[4]

A child of the Enlightenment, the French Revolution was supposed to inaugurate a new Age of Reason. As a symbol essentially representing its true spirit,[5] the new calendar was thus expected by its architects to help promote clarity and precision and substitute "the reality of reason for the visions of ignorance."[6] It was by no mere coincidence that some of the eminent scholarly authorities they consulted—such as the mathematicians Joseph Louis Lagrange and Gaspard Monge—were also members of the committee that had just introduced the metric reform of the traditional system of measures and weights. The French Republican calendrical reform was essentially an extension of the latter, and the day was to become the functional analogue of the meter and the gram. The new system of units of time was largely based on the decimal principle, which is one of the cornerstones of Western mathematics. Along with the introduction of the new ten-day week, the reformers divided the day into ten hours, each of those into 100 "decimal minutes," and each of the latter into 100 "decimal seconds," so that practically all the new units of time shorter than the month were interrelated in decimal terms! The architects of the reform were fully aware of the tremendous symbolic significance of that. When trying to legitimize the introduction of the *décade,* they were careful to emphasize the "clear reason" underlying the "rational" and "scientific" decimal system, particularly in contrast to the supposedly superstitious astrological basis of the seven-day week.[7]

The real target of the reform campaign, however, was the Christian, rather than the astrological, seven-day week, and, from a symbolic standpoint, the abolition of the seven-day "beat" expressed the wish to de-Christianize France far more than the attempt to make life there more "rational." Whereas the traditional calendar had been associated with the priesthood and with "Catholic superstition," the new calendar was supposed to be a civil calendar, divested of any religious associations.[8] As Maréchal had originally proposed, "the calendar of the French Republic . . . must not resemble in any respect the official annuals of the apostolic and Roman Church."[9] It is, therefore, hardly surprising that his original reform proposal, his *Almanach des Honnêtes Gens,* had actually been burnt by the Bourbon government as "impious, sacrilegious, blasphemous, and tending to destroy religion."[10]

Hence the particular symbolic significance of the abolition of the Saints' Days, the replacement of the Christian Era by a Republican Era that began in 1792, and the substitution of September 22 for January 1 as New Year's Day.[11] Yet the most significant calendrical contribution to the attempt to de-Christianize France was undoubtedly the obliteration of the seven-day week and, along with it, Sunday. Thus, when the chief architect of the new calendar, Charles-Gilbert Romme, was asked what the main purpose of the new calendar was, he could reply unequivocally: "To abolish Sunday."[12] The *décade*—or, rather, to be more precise, its "peak day," Décadi—came to be the single most important symbol of the de-Christianization of France. The Kulturkampf waging in France during the 1790s was thus largely a struggle between "Dominicans" (from *dies dominica,* the Latin for "Sunday") and "Decadists,"[13] and, indeed, it was often portrayed by the pamphleteers of that period as a struggle between Monsieur Dimanche ("Mr. Sunday," representing the Church) and Citoyen Décadi ("Citizen Décadi," representing the State).[14]

During the period commonly known as the Reign of Terror, the French Republic made great efforts to obliterate the seven-day rhythm, which was associated symbolically as well as practically with church-attending practices. Churches were allowed to open only on Décadi, and citizens were forbidden to close their stores on Sunday and wear their *habits du dimanche* ("Sunday best").[15] However, while systematically destroying the traditional seven-day rhythm, the revolutionary authorities were also busy trying to have it replaced by an alternative weekly rhythm, based entirely on the *décade.* As soon as the Revolutionary calendar was put into effect, they introduced a new set of civil holidays into French public life. These holidays were all based on this new weekly rhythm, and were appropriately called "decadal festivals" (*fêtes décadaires*).

Some of these holidays were already celebrated in late 1793,[16] yet the idea that every Décadi would be celebrated is associated with the rise of Maximilien Robespierre to absolute power in 1794. On April 6 (a day after the introduction of a new civil–religious cult was first recommended by the Committee of Public Safety, and also the day on which his archrival Danton was executed), he proposed that decadal festivals, to be celebrated on successive Décadis, be established on a regular basis.[17] Then, on May 7, he issued a decree introducing thirty-six decadal festivals—corresponding to the thirty-six Décadis of the new calendar year—each of which was to be dedicated to some abstract idea (such as patriotism or filial piety).[18] The celebration of the Festival of the Supreme Being throughout France on June 8, 1794—with Robespierre himself presiding over the festivities in the gardens of the Tuileries—was supposed to inaugurate this new annual cycle of weekly festivals, the observance of which was to be based entirely on the new ten-day rhythm.

As a functional substitute for the Church's Lord's Day, Décadi was essentially part of a "decadal religion,"[19] and was to be celebrated in holy temples (which later would indeed be called "decadal temples"[20]), opening and closing with the singing of hymns.[21] The only significant difference between this new ten-day weekly cult and the seven-day weekly cult which it was obviously supposed to replace was that it was to be consecrated to the French Republic rather than to Christ. Thus, for example, the hymn that would close the celebration of Décadi was the patriotic *l'hymne des Marseillais*, which would later become the French national anthem.

It took four years, however, before Robespierre's dream actually came true. On July 27, 1794, only seven weeks after the celebration of the Festival of the Supreme Being, he was overthrown and executed, his downfall marking the end of the Reign of Terror and the beginning of a relatively moderate three-year period. Some of the festivals being observed in France during that period—especially after the enactment of the Law of National Festivals on October 25, 1795—were still celebrated on Décadi.[22] Particularly noteworthy are the six so-called "moral festivals"—dedicated to Youth, Spouses, Gratitude, Agriculture, Liberty, and the Old, respectively—which were observed on the first Décadi of each of the last six months of the Republican calendar year. However, major annual holidays such as the commemorative anniversaries of the execution of King Louis XVI, the destruction of the Bastille, the storming of the Tuileries, the foundation of the French Republic, and the fall of Robespierre were all fixed on annual dates that never coincided with Décadi.[23] (The permanent correspondence between particular annual dates and particular days of the *décade* will be discussed later.) With not all Décadis being observed as national holidays, as Robespierre had envisioned it, the significance of the new weekly rhythm was obviously declining.

All that changed dramatically soon after the coup d'état of September 4, 1797, when the ruling Directory essentially reinstated the 1793–94 de-Christianization policy, among the major victims of which were Sunday and the seven-day week. The ten-day *décade*, originally proposed in 1788 and officially introduced in 1793, was to reach its heyday during 1798 and 1799, when it came to be at the very center of a major cult that was actually even named after it.

Part of an attempt to establish a rationalistic national "church" similar to the one envisioned by Robespierre, the so-called "decadal cult" (*culte décadaire*) was a product of Theophilanthropy, a civil religious movement that flourished among the Republican intelligentsia and the Parisian bourgeois elite and was actually patronized by one of the most influential members of the Directory, La Révellière-Lépeaux.[24] Yet the man who ought to get the full credit—or blame—for introducing and implementing it was Merlin de Douai, who actually presided over the Directory and was personally responsible for authoring the decree of April 3, 1798, which, for the first time, made the observance of the ten-day week mandatory.[25]

The decree—later ratified through the laws of August 4 and September 9[26]—constituted the first rigorous attempt ever to obliterate the seven-day week through the enforcement of the use of an alternative weekly cycle. To be sure, during the Reign of Terror, the authorities compiled lists of heads of families who did not participate in the decadal festivals, and Sunday-observers were condemned as non-Republicans as well as dangerous enemies of liberty, equality, and the poor.[27] However, legal sanctions against those who defied the *décade* were usually not enforced. That was to change dramatically in 1798, when severe fines and even jail sentences were applied to such violations as opening one's store on Décadi or closing it on Sundays that did not coincide with Décadi. The laws of 1798 made the closing of all stores, government offices and tribunals, and public as well as private schools on Décadi (as well as on Quintidi afternoon) mandatory. The dominance of the new weekly rhythm was also felt in commerce, as all fairs and markets were fixed on particular days of the *décade*. The laws also prohibited the use of the traditional designations of the days of the week in journals, contracts, and posters.

The Directory's main goal was obvious—to pull the entire social and economic life of France outside the sphere of the traditional Christian weekly rhythm, so as to make the latter absolutely irrelevant to daily life. Just as we would find it most difficult—as we shall see later—to adhere to a ten-day rhythmic pattern of activity in a social world dominated by the ubiquitous seven-day "beat," so would the French find it almost impossible to even keep track of the days of the seven-day week when almost their entire affairs would be regulated by a ten-day rhythm of activity. Furthermore, how would anyone be able to preserve the traditional Christian way of life and attend church regularly every Sunday, when stores could be closed only on Décadis and Quintidi afternoons? Similarly, given that fish markets were held only on Duodi, Quintidi, and Septidi,[28] how would citizens be able to keep eating fish every Friday?

The manner in which French citizens were supposed to celebrate the decadal festivals, which began to reappear at least since January 1798,[29] was formally spelled out in yet a third law, which the Directory passed on August 30, 1798.[30] Essentially reintroducing rites that had been practiced back in 1794 yet which had never been enforced as mandatory,[31] this law revolved around the "decadal reunion" (*réunion décadaire*), a patriotic celebration of the French Republic, that was to take place regularly every Décadi, as Robespierre had envisioned it. Since the new weekly rhythm associated with the cult of France was essentially meant to be the functional substitute for the traditional weekly rhythm associated with the cult of Christ, some parallels between the weekly celebration of the French Republic every Décadi and the traditional weekly celebration of the Lord's Day every Sunday might be expected. And, indeed, as we learn from actual descriptions of decadal reunions held during 1798 and 1799,[32] they were essentially modeled after the traditional Sunday gatherings of the Church, which they were obviously designed to replace.

From August 30, 1798, all wedding and adoption ceremonies, as well as all official announcements of births, deaths, and divorces, had to take place at the decadal reunions, and were thus tied to the new weekly rhythm of French collective life. At those reunions, magistrates would also deliver sermonlike moral lectures on citizenship and read to their communities the recent news as well as the laws that were passed during the preceding *décade*. These, along with the major themes emphasized in the moral lectures, would normally be provided in circulars which supplemented the periodic "breviary" of the decadal cult, namely the *Bulletin décadaire des affaires générales de la République*. Those circulars were issued regularly by the Minister of the Interior, François de Neufchâteau, whose particular sensitivity to minute details such as music and decor[33] makes him worthy of being remembered as the actual producer, as well as "director," of the decadal reunions.

While the *décade* and Décadi were being praised in popular songs since 1793,[34] they were also the obvious targets of satire. Thus, for example, in the 1796 Parisian comedy *Nicodème à Paris, ou la décade et le dimanche* ("Nicodeme in Paris, or the Ten-Day Week and Sunday"), the struggle between the revolutionary spirit and traditionalism is aptly captured in two youngsters' dilemma whether to get married on Sunday or on Décadi.[35] Yet the defiance of the *décade* by the French people obviously involved much more than mere satire, as many continued to rest on Sunday rather than observe Décadi.[36] Note, in this regard, that, until the decree of April 3, 1798, which explicitly outlawed the practice of "double dating," even the official *Moniteur* would still parenthetically insert the traditional designation of the day of the week after the decadal one on its date line.[37] That people would need to know whether a particular Octidi was a Wednesday or a Sunday clearly seems to indicate that the seven-day week never really lost its calendrical dominance as the major rhythm regulating the collective life of the essentially traditionalistic French population.

Given all that, the Republican authorities must have regarded the 1798 laws as absolutely necessary. And yet, as they were soon to learn, implementing those laws and replacing Sunday by Décadi turned out to be next to impossible.[38] Thus, for example, many couples who would go through a civil wedding on Décadi would still also have their marriage sanctioned by a priest on Sunday. As for the mandatory Décadi rest, it was often defied by private schools run by former monks, nuns, and priests, as well as by merchants who would take off both Décadi and Sunday! (The defiance of the Décadi rest was primarily symbolic. The transition from a seven-day week to a ten-day week did not entail a reduction in the number of rest days, since the 1798 laws also allowed for a Quintidi afternoon rest.) All in all, with the main exception of Paris and the department of Yonne, the *décade* proved to be a complete failure and, particularly among the rural population, never managed to replace the seven-day week.

Bishop Henri Grégoire was right when he prophetically told the calendrical reformers back in 1793, "Sunday has existed before you, and it will survive you."[39] Long before its official discontinuation, the decadal cult was already dying, and, following Merlin de Douai's and La Révellière-Lépeaux's resignation from the Directory on June 18, 1799, it all but completely disappeared in many parts of France.[40] However, as a child of the First Republic, the *décade* was also destined to die with it, and, at least officially, it would survive until the appearance of Napoleon Bonaparte on the French national scene.

Soon after Napoleon's coup d'état on November 9, 1799, the *Bulletin décadaire* was discontinued. Then, on December 8, the new Minister of the Interior, Pierre Simon Laplace, annulled the decree that had forced "decadal temples" to close on any day other than Décadi, thus essentially allowing them to resume functioning as churches.[41] An announcement made three weeks later by Napoleon's brother, Lucien Bonaparte, indicated that the anniversaries of the foundation of the Republic and of the destruction of the Bastille were to be the only festivals involving mandatory observance.[42] The indirect implication was that the observance of Décadi was no longer mandatory, which soon led most of the French population living outside Paris to abandon the decadal festivals altogether.[43] Then, on July 26, 1800, Napoleon and his two assisting consuls issued a decree announcing that, with the single exception of public officials, who would still be bound by the mandatory Décadi rest, French citizens were free to rest on whatever days they wished.[44] On the following Décadi, only half of Paris's storekeepers kept their stores closed, and many of them were already condemned publicly as "Jacobins."

Following that, the fixing of markets and fairs on particular days of the *décade* and the fact that marriages were still being considered valid only from Décadi were the only significant remaining traces of the ten-day week in the life of French citizens other than public officials. People soon abandoned the decadal rhythm altogether, and on April 18, 1802 even the *Moniteur* went back to using the traditional designation of the day of the week on its date line.[45] Finally, on September 9, 1805, the official Sunday rest—along with the Gregorian calendar—was legally reinstated, and the restoration of the seven-day "beat" was completed.

The seven-day week was restored only after the conclusion of the concordat between Napoleon and the Pope, which essentially reestablished the Church in France. Just as introducing the *décade* was part of a general attempt to de-Christianize France, the restoration of the seven-day week was an integral part of Napoleon's general policy of reconciliation with the Church. The failure of the decadal experiment must therefore be understood within the context of the overall failure of the Revolution to de-Christianize France.

The introduction of the *décade* was undoubtedly one of the boldest attempts in history to obliterate the seven-day week, and the years 1793–1805 were definitely the darkest days of the latter since having been introduced to the West. The complete failure of this most impressive calendrical adventure ought to teach us a striking lesson about the tremendous resilience of tradition in general and of religion in particular. To further appreciate that lesson, we shall now examine an even more radical attempt, made some 140 years later in the Soviet Union, to destroy the seven-day week.

The Soviet Five-Day *Nepreryvka*

In May 1929, at the Fifth Congress of the Soviets of the Union, a major reform of the existing workweek, which would involve the introduction of a so-called "uninterrupted production week," was proposed by delegate Larin.[46] The proposal attracted relatively little attention at the congress itself, yet Larin soon managed to get Joseph Stalin interested in it, and, within a couple of weeks, the Soviet press was already raving about his plan. By June, when Larin's proposal was examined by the "Rationalization Section" of the Supreme Economic Council, the Commissar of Labor, Ouglanov, was already its only remaining significant opponent, and, by late July, any opposition to the uninterrupted production week was ideologically crushed as "counterrevolutionary bureaucratic sabotage." Finally, on August 26, 1929, the Council of People's Commissaries of the Soviet Union officially announced that, starting from October 1, a major transition of all productive enterprises as well as offices from the traditional interrupted workweek to a continuous production week would be put into effect.[47]

A year earlier, when the Soviet government launched its first "five-year plan," a socialist program of speedy industrialization, maximizing output growth was clearly one of the top items on its agenda. That obviously presupposed exploiting industrial equipment to its utmost, and the authorities resolved to reduce waste by making sure that the expensive machines would be utilized incessantly and never stand idle. The introduction of a continuous working day, based on a multiple-shift system that allowed production to proceed in an uninterrupted fashion even during nighttime, was an obvious product of this new Soviet emphasis on the continuity of industrial production. The traditional workweek, however, still involved an unproductive, wasteful weekly day of rest on which the output of the expensive equipment was precisely zero. Hence the various attempts, since 1927, to experiment with a *continuous workweek*.[48] These experiments clearly led to the introduction of the *nepreryvka* ("uninterrupted") in October 1929.

Maximizing output growth must have been Larin's main objective, as one can tell from the title of his original reform proposal, "Three Hundred or Three Hundred and Sixty,"[49] which obviously alluded to the prospect of exploiting industrial equipment sixty more days every year. It was also the main rationale provided by the Council of People's Commissaries of the Soviet Union in their decree of August 26, 1929. And yet, if the concern with economic waste were indeed the only factor motivating the calendrical reform, the Soviet authorities could have easily preserved the seven-day week and simply replaced the traditional common weekly day of rest by seven different rest days all staggered vis-à-vis one another, just like the various shifts in the multiple-shift system. However, the introduction of a continuous workweek turned out to be only one component of an essentially twofold reform of the week. On September 24, 1929, a week before the *nepreryvka* was put into effect, the Council of People's Commissaries of the Soviet Union modified its original decree of August, adding that the new workweek would be a five-day, rather than a seven-day, cycle, with workers resting every fifth, rather than every seventh, day.[50]

To appreciate the antireligious significance of this move, note that, originally, the days of the week were even supposed to retain their traditional names, with only Saturday and Sunday being removed from the weekly cycle.[51] A satirical cartoon published in a Russian émigrés' newspaper from that period shows the same two days being shot by a Soviet soldier "for their bourgeois origins."[52] These two weekly bastions of Judeo-Christian religious sentiments were clearly the main targets of a regime vehemently determined to fulfill the Marxist dream of crushing the "opiate of the masses." In fact, when the Commissar of Labor expressed his concern about the future of Sunday, he was told explicitly that the calendrical reform was introduced essentially to "combat the religious spirit."[53]

As in France 140 years earlier, the main purpose of abolishing the seven-day week in the Soviet Union was to destroy religion there.[54] (Interestingly enough, shortly after the 1917 Revolution, a few attempts were made to follow the spirit of the French Revolution and introduce the ten-day *décade* to the Soviet Union.[55]) Altering the length of the weekly cycle was supposed to pull the entire social and economic life of the Soviet Union outside the sphere of relevance of the traditional seven-day rhythm associated with its three major religions (Christianity, Islam, and Judaism), so as to make that rhythm of no use for any purpose whatsoever (and, thus, both obsolete and dispensable). In a social world where one's most important affairs would all be regulated in accordance with a five-day rhythm of activity, it would be most difficult to keep track of the traditional seven-day cycle and not to lose count of one's days of religious worship. More significantly, only once every thirty-five days, when the traditional and new weeks would coincide, would a Soviet worker be able to actually attend church on Sunday, mosque on Friday, or synagogue on Saturday. Thus, on any given traditional weekly day of public worship, only one fifth of the entire Soviet work force (and not the same people every week) would be able to attend services, the other 80 percent being at work!

The days of the new weekly cycle were originally supposed to either retain their traditional Monday-through-Friday names or assume "revolutionary" names such as Trade Union, Soviet, Lenin, Komsomol, Party, Hammer, and Sickle, yet very soon they came to be known simply as "first day," "second day," and so on.[56] However, on some calendars as well as on slips indicating to workers the days on which they were off duty, particular days of the week soon also came to be associated with particular colors—the first day with yellow, the second with peach, the third with red, the fourth with purple, and the fifth one with green. It was also not uncommon that, in address books, people would add to the names of friends and acquaintances the color corresponding to the day of the week on which they were off duty.[57]

The considerable salience of days off must be understood within the context of the peculiar temporal organization of the Soviet society from October 1929. It was quite understandable that one would become associated with one's regular weekly day off work, since the latter would be shared by only one fifth of the rest of the Soviet society. Soviet workers may have rested more often than their Western counterparts (once every five, rather than seven, days), yet they certainly did not rest together, as one society, since 80 percent of the entire Soviet working population would be at work on any given day.

In order to guarantee the continuous operation of productive enterprises, the Soviet authorities made a calendrical experiment that was essentially much more radical than the French reform of the week. Rather than merely alter the length of the week from a seven-day cycle to a five-day one, they essentially tried to destroy the idea of a common societal weekly cycle by abolishing the traditional Judeo-Christian institution of a single, uniform weekly day of rest that is commonly shared by the entire society.

Within Judaism, Christianity, or Islam, a single weekly rhythm also involves a single weekly cycle. However, a brief glance at the relations among those religions reminds us that such a state of affairs should not be taken for granted. Even within the Soviet Union itself, Christians, Moslems, and Jews had always adhered to the very same seven-day weekly rhythm while, at the same time, living in accordance with three distinct weekly cycles of activity that would peak on Sunday, Friday, and Saturday, respectively. (Likewise, within contemporary American society, the fact that many restaurant and museum employees are normally off on Mondays ought to remind us that a seven-day work/rest "beat" does not necessarily mean having a regular day off on Sundays.)

The main social theme underlying the introduction of the *nepreryvka* was the obliteration of *temporal symmetry*, a traditional form of coordination that involves synchronizing the activities of different individuals so that they would take place together.[58] The Soviet reformers essentially tried to replace temporal symmetry with *temporal complementarity*,[59] an alternative pattern of coordination which involved staggering the activities of Soviet social life.[60] They replaced the seven-day week by no less than five new weeks, which, despite being of the same length, were nevertheless separate cycles that revolved around five different weekly days of rest. In short, by introducing the *nepreryvka*, the Soviet authorities essentially divided the entire society into five separate working populations, staggered vis-à-vis one another like the different voices in a polyphonic, five-voice fugue!

The "togetherness" brought about by temporal symmetry clearly enhances social solidarity, and Soviet workers who shared the same days off work were tied to one another by a special bond, quite like the one that exists among night-shift workers.[61] That one would tend to choose one's friends from among those who shared the same days off is quite understandable, particularly given that, on days when workers had a day off, only about 20 percent of the people they knew would be available for socializing, the other 80 percent being at work. The problems inherent to trying to get together with people whose work schedules are essentially out of phase with one's own are obvious.

However, as indicated by the following complaint, which appeared in the official newspaper *Pravda* on the very day the *nepreryvka* was put into effect, such problems were dwarfed by the actual disruption of family life brought about by the reform: "What is there for us to do at home if our wives are in the factory, our children at school, and nobody can visit us. . . ? It is no holiday if you have to have it alone."[62] Some degree of temporal symmetry is necessary for maintaining healthy family relationships, and the new emphasis on temporal complementarity obviously disrupted family life in the Soviet Union. The impossibility of bringing an entire family together other than after a long day's work or on the few annual holidays must have contributed considerably to the erosion of Soviet family life in this period.

Given the traditional Marxist aversion toward the family, it is quite conceivable that the eventual destruction of the family may have even been on the actual agenda of the architects of the Soviet calendrical reform.[63] Lenin's widow Nadia Krupskaya, for example, explicitly regarded Sunday family reunions as a good enough reason for abolishing that day.[64] However, even if that had indeed been the case, workers' widespread discontent soon led the authorities to reconsider the matter, and, on March 16, 1930, the "Government Commission of the Council of Labor and Defense on the Transition of Enterprises and Offices to a Continuous Production Week" began recognizing families' requests for synchronized days off work as a legitimate factor to be considered upon designing work schedules.[65]

Thus, in introducing the *nepreryvka*, the Soviet authorities were probably motivated not only by their wish to maximize output growth, but also by their aversion toward religion as well as the family. One other motivating factor may have been their inclination toward the *divide et impera* ("divide and rule") form of political dominance. The fact that only 20 percent of the entire Soviet work force would share a day off together on any given day probably made it most difficult for any serious political organization to get off the ground.

Along similar lines, the institutionalized absence of one fifth of the work force from work on any given day also ruled out the possibility of ever having general workers' meetings (which, ironically, had always been regarded by Marxists as necessary for the development of a strong class consciousness among the proletariat). The fact that, on any given day, only 80 percent of the workers would be at work also caused considerable problems in the management of work itself.[66] Most disruptive, however, were the problems of having to organize replacements for workers on the days they were off, and of maintaining the continuity of work despite the obvious discontinuities created daily by workers leaving for, or coming back from, their weekly days off. (Note also the peculiar situation of returning from one's day off work and having to catch up with one's fellow workers, who had been working in the meantime, a situation which most of us, who rest on the very same days that our fellow workers do, normally do not confront.)

The Soviet authorities recognized these problems and tried to address them.[67] In an attempt to alleviate the problem of continuity, they introduced special "transition periods"[68] designed to allow for a relatively gradual process of picking up as well as handing over responsibilities. Workers occupying highly specialized or executive functions were allowed to take off only the second or the fourth day of the new week. The third day could thus serve as a sort of "bridge" on which those who were just about to leave for their "weekend" would brief those who had just returned from theirs on current developments at work. Also, all important meetings were scheduled only for the first, third, and fifth days of the week, so that those specialists and executives would never have to miss any of them. All other workers were divided into five roughly equivalent groups, each of which would rest on a different day of the week. Industrial plants often also employed "flying squads," consisting of supplementary workers with special "transferable" skills that allowed them to switch rapidly from one type of function to another. All these devices were part of a general attempt to depersonalize occupational functions so as to make more workers easily replaceable on a regular weekly basis. In facilitating the interchangeability and substitutability of workers, impersonality is probably the most distinctive characteristic of all bureaucratic systems.[69] Its particular indispensability for institutions that operate on a continuous basis[70] accounts for its special significance for the organization of Soviet labor after the introduction of the continuous production week.

One of the dangers of impersonalizing occupational functions, however, is the phenomenon of "floating responsibility,"[71] whereby, when a particular responsibility may be assumed by more than one person, it is very often not assumed at all. And indeed, following the introduction of the *nepreryvka*, when workers were being replaced regularly by some other workers on one out of every five days and would thus no longer assume full responsibility for any task or equipment, there was a sharp decline in personal responsibility among Soviet labor. As early as spring 1930, even official organs were already reporting slower work and worse service, and attributing these to the continuous production week.[72]

The connection between irresponsibility and impersonality (both of which are denoted in Russian by one and the same word, *obeslichka*) is made explicit in a 1931 cartoon, captioned "Go away, a pass without a personal picture is invalid," and portraying irresponsibility as a faceless old woman being chased out of an industrial plant.[73] The cartoon was published very shortly after Stalin's speech on the Soviet economic policy before a conference of business managers and industrial administrators on June 23, 1931. In that speech, Stalin singled out irresponsibility as the most urgent problem and the worst enemy that had "crept into our enterprises as an illegitimate companion of the continuous workweek":

> many of our enterprises went over to continuous production too hastily, without preparing the necessary conditions, without properly organizing the shifts, so that they should compare more or less favorably in skill, without establishing the responsibility of each worker for a given task. As a result of this the continuous workweek, left to take its natural course, has led to irresponsibility. . . . As a result we have the lack of any sense of responsibility for work, careless handling of machines, mass breakage, and the absence of an incentive to increase the productiv-

ity of labor. . . . It follows from this that some of our comrades have been in too great a hurry in some places in introducing the continuous work week, and in their haste perverted the continuous work week by transforming it into a reign of irresponsibility.[74]

Claiming that "it would, however, be incorrect to say that the continuous working week inevitably leads to irresponsibility in production," Stalin was clearly not ready to admit that the idea of introducing a continuous workweek might have been a mistake. However, being unable to figure out how to pour out the water without also spilling the baby, he recommended abandoning the *nepreryvka* altogether. He promised to restore "a real continuous work week without irresponsibility" when Soviet industry would be more adequately prepared for its successful reintroduction. In fact, as late as 1933, there were still some who seriously believed that the discontinuation of the *nepreryvka*—which had already been adopted by almost the entire Soviet industry[75]—was only temporary.[76] As it turned out, however, the decree issued by the Council of People's Commissaries of the Soviet Union on November 23, 1931[77] came to mark the actual death of that calendrical experiment, only slightly more than two years after its inception. Following that decree, with the exception of a few instances where it was retained for several more years (for example, in public transportation, cooperative stores, and dining halls), the *nepreryvka* soon disappeared from the pages of history forever.

The discontinuation of the five-day week, however, by no means marked the restoration of the seven-day week. As we have already seen, more than just economic considerations had been responsible for the Soviet reform, and, indeed, as soon as the economic rationale for the abolition of the seven-day week disappeared, the much deeper antireligious sentiments of the reformers surfaced. The new cycle that came to replace the *nepreryvka* reinstated temporal symmetry through the reestablishment of a single weekly day of rest that was commonly shared by the entire Soviet society as a whole. However, while no longer being derived from a five-day weekly rhythm, neither was it based on the traditional seven-day rhythm associated with religion. From December 1, 1931, with very few exceptions, work throughout the Soviet Union was structured in accordance with a new, six-day week, the *chestidnevki*.[78]

Between December 1931 and June 1940, every sixth day was regarded in the Soviet Union as a common day of rest. (To compensate workers for the considerable reduction in the number of rest days, the number of daily working hours was reduced accordingly.[79]) As the reformed Soviet calendar year consisted of thirty-day months, these rest days were permanently fixed on the sixth, twelfth, eighteenth, twenty-fourth, and thirtieth days of each month. The restoration of the Gregorian calendar, however, reintroduced seven 31-day months and one 28-day month that could no longer be "neatly" subdivided into precisely five 6-day weeks. As no uniform arrangement was officially implemented, most offices and industrial plants chose to close on both the thirtieth and thirty-first days of thirty-one-day months. As for the end of February—some of them would close on March 1, thus allowing for two 4-day workweeks between February 24 and March 6, whereas others would operate continuously for nine days (or ten, on leap years) between those two days.[80]

While many Soviet citizens were rapidly losing count of the days of the seven-day week,[81] even leading official organs such as *Pravda* did not seem to be able to ignore the traditional cycle and found it necessary to keep printing its days on their mastheads.[82] As in France 140 years earlier, it was the essentially traditionalistic rural population who spearheaded the movement to preserve the seven-day week. When the authorities insisted that they rest in accordance with the new secular weekly rhythm, many peasants followed the example set by their French predecessors and sabotaged their efforts by taking off both the official rest days and their traditional weekly days of worship, which they defiantly marked on the official calendars issued by the government's printers. Economic life in the countryside was also still tied to the traditional seven-day rhythm, and, throughout the 1930s, Saturday nights could be easily recognized by the heavy traffic of peasants' carts packing the roads on their way to the traditional markets, which were still being held regularly on Sundays. The authorities soon acknowledged their failure to defeat the seven-day week in the countryside when they decided to fix election days on official rest days which also coincided with Sunday (or, in the case of the predominantly Moslem republics, with Friday).[83]

The Soviet calendrical adventure finally came to an end on June 26, 1940, when the Presidium of the Supreme Soviet abolished the *chestidnevki* and restored the seven-day week.[84] The official rationale offered for the prolongation of the weekly cycle was, not surprisingly, the need to increase production. A closer examination, however, reveals that economic considerations, which had played only one part in the decision to abolish the seven-day week in 1929, were also only partly responsible for its restoration in 1940.

Given the peasants' success in resisting the government's attempt to secularize the week, the Soviet Union was slowly becoming two distinct societies, completely out of phase and temporarlly uncoordinated with one another—the city living in accordance with the official civil six-day rhythm, and the country stubbornly sticking to the traditional religious seven-day cycle. The coexistence of these two conflicting weekly rhythms obviously eroded social solidarity at the societal level, aside from being a most conspicuous testimony of the government's failure to assert its authority.

To appreciate the role played by religious sentiments in the downfall of the *chestidnevki,* note that, along with their restoration of the seven-day weekly cycle, the Soviet authorities also reestablished Sunday as the official weekly day of rest. Had it not been for powerful religious pressures which it apparently could not resist, the government could have easily chosen any of the other six days of the restored cycle, particularly given the fact that Sunday—the religious associations of which resonate even in its name, Voskresen'e, which literally means "Resurrection"—had officially been dead for almost eleven years!

The complete failure of the eleven-year Soviet calendrical experiment, just like that of its French predecessor 140 years earlier, attests to the tremendous resilience of tradition in general and of religion in particular. In both France and the Soviet Union, some desperate attempts were made by two of the most ruthless totalitarian regimes in history to completely destroy the Judeo-Christian seven-day week. In both societies, to this day, it still remains the dominant "beat" of social life.

CHAPTER THREE

Cultural Variations on a Theme

T HE COMPLETE FAILURE of the French and Soviet calendar reforms might lead one to regard the seven-day week as inevitable. That, however, would be an erroneous conclusion. It is only the tremendous resilience of religious tradition that enabled the seven-day week to withstand the fierce onslaughts of the ten-day, five-day, and six-day weeks. The French and Soviet experiments failed only because they were imposed on societies deeply committed for many generations to the seven-day religious "beat." However, in the absence of such prior traditional-religious commitments, weekly cycles that are not seven days long can very well be established and survive for centuries.

Equating the institution of the week with a seven-day rhythm is the consequence of an ethnocentric bias which can be challenged by the surprisingly wide variability of the week's length in different parts of the world as well as in different historical periods. The examples that follow also demonstrate the variability of the week's functional use, from the regulation of religious and economic activity to the construction of divinatory calendars. All of them, however, involve one and the same phenomenon, namely, a regular, continuous cycle that is longer than the day yet shorter than the month.

The Market Week

The evolution of the week generally coincided with the rise of a market economy,[1] and it is, therefore, hardly surprising that the regulation of economic transactions was one of the earliest functions of this cycle. Periodic markets are most characteristic of peasant economies, where aggregate demand is not sufficient for supporting permanent shops,[2] and, to this day, the market week still flourishes in developing countries around the world.[3] The three-day market weeks of ancient Colombia and New Guinea, the five-day market weeks of ancient Mesoamerica and Indochina, and the ten-day market week of ancient Peru all serve to remind us that such *weekly market cycles* have not always been seven days long.[4]

The ancient southern Chinese twelve-day week is a classic example of a weekly cycle that served to regulate economic transactions.[5] Three-day market cycles—regularly held on the first, fourth, seventh, and tenth days of the week—were clearly derived from it. So were the six six-day market cycles, which were regularly held on the first and seventh, second and eighth, third and ninth, fourth and tenth, fifth and eleventh, and sixth and twelfth days of the week, respectively. These six market cycles played such a major role in the economic life of ancient China that they were even assigned distinctive names—Tzu-wu, Ch'ou-wei, Yin-shen, Mao-yu, Ch'en-hsü, and Ssu-hai.

Sometime around the eighth or seventh century B.C., an eight-day market week evolved in the area presently known as Italy. Part of an Etruscan time-reckoning system based on the number 8, it essentially revolved around a *periodic market day* that was held regularly every eight days.[6] The Romans inherited this weekly cycle from the Etruscans no later than the early sixth century B.C., and, for about seven or eight centuries, much of the social and economic life of Rome was organized in accordance with its "beat."

The Roman eight-day week was known as *internundinum tempus* or "the period between ninth-day affairs." (This term must be understood within the context of the ancient Roman mathematical practice of inclusive counting, whereby the first day of a cycle would also be counted as the last day of the preceding cycle.[7]) The "ninth-day affair" around which this week revolved was the *nundinae*, a periodic market day that was held regularly every eight days. On that day, farmers would come to the city to sell their products. However, the nature of the Roman week was by no means purely economic, as the periodic contact between country and city obviously involved much more than just the transactions at the marketplace. Much of

Rome's social life essentially revolved around the *nundinae*. On that day, all schools and courts were closed, no public meeting was held, and Romans would stop working, go to the public bath, and meet their friends at banquets. On the market day, Romans were also allowed to host five country guests instead of the regular maximum of three.[8]

The decline of the eight-day week coincided with the expansion of Rome.[9] A periodic market cycle could flourish only at a time when Rome was still a relatively small city. Unlike peasant economies, urban economies necessitate permanent, continuous commerce, so that, as Rome was expanding, the periodic commerce between farmers and burghers was becoming obsolete. Coincidentally, the astrological and Christian seven-day weeks that had just been introduced into Rome were also becoming increasingly popular. There is evidence indicating that the Roman eight-day week and those two seven-day cycles were used simultaneously for some time.[10] However, the coexistence of two weekly rhythms that were entirely out of phase with one another obviously could not be sustained for long. One of them clearly had to give way. As we all know, it was the eight-day week that soon disappeared from the pages of history forever.

Interestingly enough, two of the Roman inscriptions that point to the historical overlap between the declining eight-day market week and the rising seven-day astrological week during the early days of the Empire also seem to indicate that the former was assuming a more complex form. In one inscription, for example, eight different communities are listed together in a special column under the heading *"nundinae"*—Pompeii, Nuceria, Atilla, Nola, Cumae, Putiolae, Rome, and Capua.[11] Rome was obviously still holding a market every eight days, yet the weekly market cycle was now connecting eight different cities that were holding eight different, albeit coordinated, markets on successive days of the week. This peculiar phenomenon can probably best be explored within an entirely different context, namely, contemporary West Africa.

Due to Christianity and Islam, the seven-day week has become an integral part of the current African scene. A lot of economic activity throughout that continent, however, is still organized in accordance with various indigenous market cycles that are not seven days long. Quite popular in Rwanda, Tanzania, Cameroon, Togo, and Zaire only a few decades ago, three-day, five-day, six-day, nine-day, and ten-day market cycles still regulate the economic life of various tribes in Ghana, Nigeria, and the Upper Volta. The most popular of all indigenous African weeks, however, is the four-day market week, along with the eight-day and sixteen-day weekly market cycles that have most probably derived from it. That week has been used in parts of West Africa, East Africa, and the lower Congo at least since the late seventeenth century. Today it prevails throughout a long continuous belt stretching from eastern Ghana, through Togo, Benin, and Nigeria, to Cameroon.[12]

The West African week is essentially a market cycle, and its users do not always make a conceptual distinction between its temporal form and the actual economic activity it regulates. The Tiv word *kasoa* and the Efik word *urua*, for example, denote both "week" and "market." The conceptual affinity between the two is also quite evident from some of the Ewe and Yoruba names of the days of the week—"market day," "second day of the market," "market day is tomorrow," and so on.[13]

West Africans very often also do not make a conceptual distinction between the days of the weekly cycle and the places where weekly markets are being held.[14] This peculiar correspondence between space and time is a result of the fact that the West African week essentially revolves around—as well as regulates—a system of several markets that are held in more than one village. For nearly one third of the entire rural population of West Africa, who travel almost daily from one village to another in accordance with the market cycle,[15] knowing what day it is also entails knowing where a market is being held on that particular day. (Given the considerable distances among some villages, small errors in reckoning the day of the week may actually cost shoppers many miles of walking in vain to the wrong village!)

Space, time, and social structure are all interrelated in the West African market, since the system of market days that constitute a week is intimately associated with the social system of villages known as the market "ring" or "circuit."[16] The constituent members of a market ring essentially manifest their social solidarity as parts of one and the same social system through their cooperation in the establishment of a unifying weekly cycle that applies to all of them and only to them. The coexistence of several market weeks in the same area—a most conspicuous manifestation of which is the designation of any given day by more than a single name[17]—clearly indicates the coexistence of several distinct social systems there. That also explains why a rebellious village that wishes to defy the authority of its ring might express its dissent through the establishment of a new weekly market cycle that would conflict with that of its ring.[18]

One of the major manifestations of the interdependence among parts of whole social systems is *temporal coordination.*[19] It was only the incorporation of otherwise isolated communities into integrated transportation systems, for example, that led to the adoption of a single, uniform standard of time throughout England and of a system of fully coordinated time zones throughout the United States.[20] Likewise, it is the temporal coordination among constituents of market rings that serves as evidence that they are essentially interdependent parts of systemic wholes. In the particular case of West Africa, this temporal coordination is manifested through the correspondence between the number of days in the market week and the number of villages in the market ring. These villages coordinate their markets by making sure that they are held regularly, in a fixed as well as continuous sequence, on successive days of the weekly market cycle. Thus, if two markets are being held in two villages on the very same day, these villages are clearly parts of two different market rings.

In other words, villages that belong to one and the same market circuit must hold their markets on different days of the week. This is designed to minimize the competition among them,[21] just as people who wish to hear one another must abide by the rules of turntaking,[22] which preclude simultaneous talk. This application of a *rotation system*, which essentially involves the establishment of coordinated differences, also resembles the way hospitals, in an effort to maintain the continuity of their patient coverage around the clock and throughout the year, arrange for nurses and physicians to work on different shifts as well as to take their vacations on different months and their nights off duty on different days of the week.[23] It is a manifestation of a particular form of temporal coordination that was examined with regard to the Soviet continuous workweek, namely temporal complementarity. The presence of temporal complementarity among West African villages that constitute market rings indicates that they are essentially interrelated through "organic solidarity,"[24] a particular form of social solidarity primarily characterized by complementary, coordinated differentiation. Given the ever-increasing division of labor in society, there is a growing presence of temporal complementarity in most modern social systems.[25] Some of the major week-related manifestations of that will be examined later.

The Baha'i Week-Calendar

Another variation of the week, a nineteen-day cycle of social and religious activity, was originally introduced in 1844 by the Persian prophet Seyyèd Ali Mohammed—better known as the Báb—who founded the Babi movement, and later adopted by the Báha'u'lláh, who, in 1863, broke away from Babism and created the international religious movement known to this day as Baha'ism. This cycle essentially revolves around the periodic observance of the Bahá (or "Nineteen-Day Feast"), a weekly reunion consisting of public worship, communal eating, and socializing.[26] Most Baha'is today live in environments organized along the conventional seven-day weekly rhythm. At the same time, however, they must also be oriented within their own nineteen-day weekly cycle, which alone regulates the spiritual as well as social aspects of their existence as Baha'is. It is quite clear that Baha'is' adherence to this peculiar weekly rhythm, to which probably no one else in the world adheres, adds considerably to their distinctive group solidarity.

The nineteen-day week is actually only one of five Baha'i units of time that are all nineteen times longer or shorter than one another—the day, the 19-day week, the 19-week year, the 19-year *vahid*, and the 361-year *kull-i-shay*.[27] The choice of 19 as the basis of the entire Baha'i time-reckoning system is congruent with the general numerological significance of this number—often considered the mystical number of unity[28]—within Baha'i tradition.[29] A classic manifestation of that is the organization of the *Béyan* (the holy Babi book attributed to the Báb) in nineteen-chapter sections (each of which, incidentally, is called *vahid*, just like the nineteen-year cycle).[30] Moreover, when completed, that work was supposed to consist of nineteen such sections, so that the total number of chapters would have been 361, "the number of all things" in accordance with which God is believed to have organized the entire universe.[31]

However, the choice of 19 as the mystical number of unity and of 361 as "the number of all things" may itself have had a calendrical origin. The special numerological significance of 19 may very well be related to the fact that it is the integer that most closely approximates 19.1113, the square root of the total number of days in the solar year, 365.2422. In other words, the Baha'is' nineteen-day weekly cycle is the closest approximation of the square root of the annual cycle. By introducing it, they have managed to establish the most symmetrical relationship possible between the week and the year, which no one else throughout history has ever managed to accomplish. This almost perfect symmetry is manifested in the unique Baha'i calendrical practice of assigning identical names—Bahá, Jalál, Jamál, 'Azamat, Núr, Rahmat, Kalimát, Kamál, Asmá', 'Izzat, Mashíyyat, 'Ilm, Qudrat, Qawl, Masá'il, Sharaf, Sultán, Mulk, and 'Alá'[32]—to the nineteen days of the week and the nineteen weeks of the year!

The unique Baha'i practice of assigning identical names to the days of the week and to the weeks of the year is made possible only by the unique symmetrical relationship between the Baha'i weekly and annual cycles. The practice of assigning distinctive names to the various weeks of the calendar year, however, is also found in two other calendrical systems, namely the Central American and Indonesian calendars. (Note also, in this regard, the traditional Jewish association of the various Sabbaths of the calendar year with the particular portions of the Pentateuch that are regularly read on them.) This practice may very well be related to one other basic structural feature that the Baha'i, Central American, and Indonesian calendars share in common, namely the absence of the unit "month." Other than the week and its precise multiples, no intermediate unit of time between the day and the year has been built into any of these three calendrical systems. These calendars are essentially based on the weekly cycle, and I refer to them as *week-calendars*. (Both the Baha'i nineteen-day week and the Central American twenty-day week are often referred to as "months." However, they bear no connection whatsoever to the lunar cycle, and, therefore, fit my definition of "week.")

The Central American Week-Calendars

The French Republican ten-day *décade* is not the only weekly cycle derived from the conventional counting system in use. The pre-Columbian civilizations of Central America developed a rather similar week. The difference between these two cycles, however, lies in the fact that, whereas the French system of numeration was decimal (that is, based on various gradations of ten), the one used by the ancient inhabitants of Mesoamerica was vigesimal (that is, based on various gradations of 20).[33] Given the centrality of the number 20 to this counting system, and the fact that "the major use to which [ancient Central American] arithmetic was put was calendrical,"[34] we might expect a twenty-day week to play a central role in the life of the ancient inhabitants of Central America. An examination of the time-measurement and calendrical systems of two of the major pre-Columbian civilizations of Mesoamerica, namely, the Maya and the Aztecs, seems to indicate that that indeed was the case.

The first manifestation of this weekly cycle is in the form of the twenty-day *uinal*, which served as the cornerstone of the entire Maya time-measurement system. Modeling the latter after their system of numeration, the Maya established a series of units of time, the length of which, in terms of number of days, corresponded to various gradations of the number 20. Consider, for example, such units as the 400-day *huna* and the 8,000-day *may*, used by Maya tribes such as the Quiche, Zuhutil, and Cakchiquel.[35] Most of the Maya, however, did not incorporate the *huna* into their conventional time-measurement system, establishing instead a rough approximation of the actual annual cycle, the 360-day *tun*, in the gradation immediately above the *uinal*. Yet that was the only interruption of a series of units of time, each of which was precisely twenty times longer or shorter than its neighboring units—the day, the 20-day *uinal*, the 360-day *tun*, the 7,200-day *katun*, the 144,000-day *baktun*, the 2,880,000-day *pictun*, the 57,600,000-day *calabtun*, the 1,152,-000,000-day *kinchiltun*, and the 23,040,000,000-day *alautun*.[36] (The only somewhat parallel system was the French Republican time-measurement system, where the week, the day, the hour, the minute, and the second were all precisely 10 or 100 times longer or shorter than one another.[37]) This time-measurement system was used for computing temporal "distances" between points in history as well as for chronological dating.[38] Thus, for example, the conventional Maya chronological dating framework (generally known as the "Long Count" or "Initial Series" system), essentially modeled after the Maya positional, place-value system of numeration, was based entirely on the above units of time. Maya dates would thus always consist of five digits which represented the number of complete *baktuns, katuns, tuns, uinals,* and days that had elapsed since some conventional "zero date" in history.

The second manifestation of the twenty-day week was within the context of the solar calendar.[39] The solar calendar year (the Maya *haab* and the Aztec *xihuitl*) was 365 days long, and, aside from five "blank" days that will be examined later, was essentially made up of eighteen 20-day weeks. (Like their Baha'i counterparts, these weeks were designated by distinctive names. The Maya called them Pop, Uo, Zip, Zotz, Tzec, Xul, Yaxkin, Mol, Ch'en, Yax, Zac, Ceh, Mac, Kankin, Muan, Pax, Kayab, and Cumhu. The Aztecs named them Atlcahualo, Tlacaxipehualiztli, Tozoztontli, Huey Tozoztli, Toxcatl, Etzalcualiztli, Tecuilhuitontli, Huey Tecuilhuitl, Tlaxochimaco, Xocotl Huetzi, Ochpaniztli, Teotleco, Tepeilhuitl, Quecholli, Panquetzaliztli, Atemoztli, Tititl, and Izcalli. Specific days within the solar calendar were designated by the name of the week to which they belonged as well as by their ordinal position within it—"6 Cumhu," "19 Atemoztli," and so on.) It was in accordance with these weeks that the Aztec *cempohuallan* market was held regularly every twenty days.[40] Furthermore, all ancient Central American festivals and religious ceremonies were celebrated in accordance with the twenty-day weekly "beat" of the solar calendar, since each one of the eighteen weeks was associated with a particular patron deity and included a distinctive festival dedicated to it. While each particular festival was celebrated on an annual basis, festivals in general occurred regularly once every twenty days and were therefore a weekly event, just like the Baha'i "Nineteen-Day Feasts," the French Republican "decadal festivals," the Jewish Sabbath, and the Christian Lord's Day.

There was yet a third context within which we can find the 20-

day week in ancient Central America, namely the 260-day divinatory calendar known as *tzolkin* among the Maya and *tonalamatl* among the Aztecs.[41] There, it was manifested in the form of a series of twenty days, each of which was associated with a particular patron deity, not unlike the days of the astrological seven-day week. The Maya called these days Imix, Ik, Akbal, Kan, Chicchan, Cimi, Manik, Lamat, Muluc, Oc, Chuen, Eb, Ben, Ix, Men, Cib, Caban, Eznab, Cauac, and Ahau. The Aztecs named them Cipactli, Ehecatl, Calli, Cuetzpallin, Coatl, Miquiztli, Mazatl, Tochtli, Atl, Itzcuintli, Ozomatli, Malinalli, Acatl, Ocelotl, Cuauhtli, Cozcaquauhtli, Ollin, Tecpatl, Quiahuitl, and Xochitl.

The 260-day divinatory calendar "year" was actually a product of the interaction between this twenty-day week and yet another weekly cycle, namely a series of thirteen days designated by numbers running from 1 through 13. The Maya associated each one of those days with one of the thirteen *oxlahunticu* ("Gods of the Upper World") who ruled the thirteen heavens.[42] At the same time, however, both they and the Aztecs also associated each entire thirteen-day week with one of twenty patron deities.[43] Each such week was thus regarded as having a distinctive character of its own (the entire Aztec week beginning with the day "1 Cipactli," for example, would be designated as unlucky[44]), and the weekly "beat" would essentially be experienced around the transition, once every thirteen days, from the dominance of one particular patron deity to that of another.[45]

The 260-day divinatory calendar "year" was essentially a *"week-year."* Totally unrelated to any natural cycle, it evolved only as a product of the interaction between two weekly cycles, namely the twenty-day and thirteen-day weeks. One might visualize that "year" in the form of two 20-cog and 13-cog wheels in gear,[46] with the latter completing twenty revolutions by the time it takes the former to complete thirteen, and with both wheels returning to the same starting position once every 260 days. As we can see in Figure 3 (where one would proceed from top to bottom and then from left to right), those two weekly cycles ran entirely independently of one another, with the names of days as well as their numbers both changing every day. The day "1 Imix," for example, would thus be followed by the day "2 Ik," rather than by either "2 Imix" or "1 Ik." The completion of any one of those weekly cycles would thus never interrupt the continuous flow of the other—"13 Ben," for example, would be followed by "1 Ix," and "7 Ahau" by "8 Imix." Since 20 and 13 have no common factor, the same combination of a particular day name and a particular day number would recur only once every 260 (that is, 20 × 13) days.

The *tzolkin* and the *tonalamatl* were essentially almanacs con-

FIGURE 3   The Maya Tzolkin Calendar

Imix	1	8	2	9	3	10	4	11	5	12	6	13	7
Ik	2	9	3	10	4	11	5	12	6	13	7	1	8
Akbal	3	10	4	11	5	12	6	13	7	1	8	2	9
Kan	4	11	5	12	6	13	7	1	8	2	9	3	10
Chicchan	5	12	6	13	7	1	8	2	9	3	10	4	11
Cimi	6	13	7	1	8	2	9	3	10	4	11	5	12
Manik	7	1	8	2	9	3	10	4	11	5	12	6	13
Lamat	8	2	9	3	10	4	11	5	12	6	13	7	1
Muluc	9	3	10	4	11	5	12	6	13	7	1	8	2
Oc	10	4	11	5	12	6	13	7	1	8	2	9	3
Chuen	11	5	12	6	13	7	1	8	2	9	3	10	4
Eb	12	6	13	7	1	8	2	9	3	10	4	11	5
Ben	13	7	1	8	2	9	3	10	4	11	5	12	6
Ix	1	8	2	9	3	10	4	11	5	12	6	13	7
Men	2	9	3	10	4	11	5	12	6	13	7	1	8
Cib	3	10	4	11	5	12	6	13	7	1	8	2	9
Caban	4	11	5	12	6	13	7	1	8	2	9	3	10
Eznab	5	12	6	13	7	1	8	2	9	3	10	4	11
Cauac	6	13	7	1	8	2	9	3	10	4	11	5	12
Ahau	7	1	8	2	9	3	10	4	11	5	12	6	13

sisting of 260 distinct types of calendar days. Each such day was a unique combination of a particular day within the thirteen-day week and a particular day within the twenty-day week. As such, not unlike the astrological hour, it had a distinctive character of its own, which was a function of the combined influences of the two particular patron deities associated with its name and number.[47] It was on the basis of its respective position within both weekly cycles that ancient Maya and Aztec soothsayers would designate any particular day as lucky or unlucky. As a matter of fact, the 260-day divinatory calendar (along with the traditional solar calendar) has been preserved to this day in many parts of rural Mexico and Guatemala, and, particularly among the Highland Maya, "day keepers" are still being consulted regularly about the most propitious days for getting married, going on trips, building houses, and launching business enterprises.[48]

The Central American solar and divinatory calendars ran entirely independently of one another, so that dates within each of these systems were reckoned quite independently of those within the other. The shortest time it would take these two 365- and 260-day cogwheels to return to the same starting position and produce the same combination of "solar" and "divinatory" dates was the 18,980-day (approximately fifty-two-year) period generally known as the "Calendar Round."[49] Consequently, there were no less than 18,980 unique combinations of "solar" and "divinatory" dates in ancient Central America.[50] Within any given fifty-two-year cycle, days would be designated in most complex forms such as the Maya date "5 Oc 16 Mol," which would designate the day Oc from the divinatory twenty-day week, which was also the fifth day within the thirteen-day week, and also occupied the sixteenth position within the "solar" twenty-day week Mol.

The above Maya date bears a superficial resemblance to a date such as "Tuesday, July 4," which designates the third day of the seven-day week which also occupies the fourth position within the seventh month of the calendar year.[51] These two dates, however, differ fundamentally in that the thirteen-day cycle as well as both "solar" and divinatory twenty-day cycles were all weeks. So were also, incidentally, the five-day market cycle,[52] and the cycle of nine nights, each of which was associated with one of the nine Maya "Gods of the Lower World" or Aztec "Lords of the Night."[53] In other words, each day in ancient Central America could be designated in terms of its positions within two distinct twenty-day cycles (the "solar" and the divinatory), a thirteen-day cycle, a five-day market cycle, and a nine-night cycle, and these were all weekly cycles!

Being based entirely on continuous cycles that are longer than the day yet shorter than the month, the traditional calendars of Central America were obviously week-calendars. Furthermore, beyond the Baha'i week-calendar, they make us realize that several different weekly rhythms can possibly all coexist within one and the same calendrical system. However, as we shall now see, it is Indonesia, rather than Mesoamerica, that presents us with the most spectacular and complex multiple week-calendar ever devised.

The Indonesian Week-Calendar

The most remarkable week-calendar ever invented evolved sometime around the ninth century on the island of Java,[54] from where it has also spread to some other Indonesian islands, such as Bali. Parts of it may have been borrowed from other civilizations—the five-day market week, for example, can also be found in Indochina, Malaya, and New Guinea,[55] and the names of the days of the seven-day week as well as the sequential order of the celestial bodies with which they are associated indicate a clear Hindu influence.[56] The entire calendar, however, is indigenously Javanese.

Various components of the Indonesian week-calendar serve various different functions. The five-day week, for example, serves to regulate market activity and, not unlike its West African counterpart, is also known by the same word that denotes "market," *pasar.*[57] The six-day, five-day, and seven-day weeks play a major role in chronological dating,[58] and it is also in terms of their combined positions within these three cycles that "annual" festivals are celebrated. As a whole, however, the Indonesian week-calendar is the functional equivalent of the Central American *tzolkin* and *tonalamatl* calendars in that its main function is divinatory. Indonesian weeks essentially form an integrated calendrical divinatory system which regulates numerous everyday activities, ranging from meeting people, shopping, and harvesting crops, to getting married or divorced, moving, launching business enterprises, going on trips, building houses, organizing puppet shows, burying the dead, and making offerings to evil spirits. Indonesians usually choose the most propitious days for performing any of those activities only after consulting special diviners, whose wooden (*tika*) or palm-leaf (*wariga*) calendars contain all the necessary information regarding the combinations of particular days from the different weekly cycles.[59]

Each such combination is particularly complex, since each day belongs to no less than nine weekly cycles! The Indonesian week-calendar is the most intricate one ever devised, because it essentially consists of nine different cycles that are all weeks. First, there is the two-day *duwiwara*, which consists of the days M'ga and P'pat. Then there are the three-day *triwara* (which consists of the days Pasah or Dora, Beteng or Waja, and Kajeng or Byantara); the four-day *tjaturwara* (which consists of the days Srí, Laba, Jaya, and Mandala); the five-day *pantjawara* (which consists of the days Umanis, Paing, Pon, Wagé, and Klion); the six-day *sadwara* (which consists of the days Tungleh, Ariang, Wurukung, Paniron, Was, and Mawulu); the seven-day *saptawara* or *uku* (which consists of the days Radite, Soma, Angara, Boda, Wĕrhaspati, Sukra, and Saneschara); the eight-day *astawara* (which consists of the days Srí, Indra, Gurú, Yama, Ludra, Brahma, Kala, and Uma); and the nine-day *sangawara* (which consists of the days Danggú, Djangur, Gigis, Nohan, Ogan, Erengan, Urungan, Tulus, and Dadí). Finally, there is the ten-day *dasawara,* which consists of the days Penita, Patí, Suka, Duka, Srí, Manú, Menusa, Eradja, Dewa, and Raksasa.[60]

The considerable intricacy of this time-reckoning system is quite evident from Figure 4, which depicts the interrelations among seven of these nine cycles, the last days of which are represented by distinctive icons. Each Indonesian calendar day is designated by nine different names because it is essentially anchored within nine different meaning contexts. As a unique combination of nine types of days from the nine weekly cycles, it is obviously assigned a unique divinatory character. The five-day and seven-day weeks are only two out of nine different weekly cycles which together produce the Indonesian divinatory calendar. And yet, even at the level of the interaction between these two cycles, the day Klion, for example, produces one type of day, Rainan, when it coincides with the day Boda, yet an entirely different type of day, Tumpĕk, when it coincides with the day Saneschara.[61]

The entire calendar "year," the 210-day *odalan,* essentially consists of 210 unique types of calendar days, each of which is defined in accordance with its relative positions within the nine weekly cycles. However, since $210 = 5 \times 6 \times 7$, the 210 types of calendar days can actually be produced by the interaction among only three of those cycles, namely the five-day, six-day, and seven-day weeks. And, indeed, all traditional Indonesian holidays, for example, are celebrated once every 210 days and are essentially defined as combinations of three particular types of days within those three weekly cycles.[62] Galungan, for example, is thus celebrated on the combined day Boda–Klion–Ariang, and Tumpĕk Pangarah on Saneschara–Klion–Tungleh. (As we can see in Figure 4, each of the thirty *ukus* of the calendar year has its own distinctive name, one implication of which is that the *sadwara* designation of holidays need not always be invoked explicitly.[63] Thus, Galungan can also be defined as the combined day Boda–Klion which falls within the *uku* Dungulan. That, however, is a mere notational matter,[64] and should not obscure the major role played by the six-day week in the establishment of the *odalan* cycle and, through that, in the regulation of "annual" holidays.)

The distinctive names of the various *ukus* of the calendar "year" are the result of a calendrical practice that can be found elsewhere only in the Baha'i and Central American calendars, and thus serve to remind us that the Indonesian calendar is essentially a week-calen-

FIGURE 4 The Indonesian Week-Calendar

Names of ukus / Days of the uku	Sinta	Landép	Ukir	Kurantil	Tolu	Gumbrég	Wariga	Warigadian	Julungwangi	Sungsang	Dungulan	Kuningan	Langkir	Médangsia	Pujut	Paang	Krulut	Mérakih	Tambir	Médangkungan	Matal	Uyé	Ménahil	Pérangbakat	Bala	Ugu	Wayang	Kélau	Dukut	Watu-Gunung
Radite																														
Soma																														
Angara																														
Boda																														
Wērhaspati																														
Sukra																														
Saneschara																														

Byantara, the last day of the three-day week

Mandala, the last day of the four-day week

Klion, the last day of five-day week

Mawulu, the last day of six-day week

Uma, the last day of the eight-day week

Dadi, the last day of the nine-day week

dar. We should remember that Indonesian dates are produced by the interaction among several cycles that are nevertheless all of the very same "level," being weeks.[65] (A temporal formulation such as Boda–Klion, for example, should be contrasted with one such as "Friday the 13th," which is the product of the interaction between two cycles of entirely different "levels," namely the week and the month.) Consider also, in this regard, the Indonesian 35-day cycle known as *tumpĕk*. On the surface, it appears to be the equivalent of our own 30-day and 31-day calendar months. Unlike them, however, it does not even begin to approximate, and has not really derived from, the actual 29.53-day lunar month, and should, therefore, not be considered a "month." Like the 42-day *adae* of the Ashanti in Ghana, which is a product of the interaction between the seven-day week and the local six-day market cycle,[66] it is essentially a product of the interaction between two weekly cycles, namely the seven-day and five-day weeks, which constitute two of the three cornerstones of the Indonesian week-calendar.

While the *tumpĕk* has nothing whatsoever to do with the lunar month, the 210-day *odalan* calendar "year" is obviously also totally disregardful of the solar year. The dissociation of the celebration of "annual" festivals and even birthdays from the solar year is a most peculiar as well as extremely rare phenomenon that is otherwise found only in the case of the Central American divinatory 260-day calendar.[67] Even in the Mohammedan lunar calendar, where festivals are celebrated once every 354 days, thus "floating" throughout the solar year, it would nevertheless take about fifteen years for a holiday to "migrate" from summer to winter. An Indonesian "annual" holiday that is celebrated this year in August, however, will next year fall in March.

That was generally also true of the Central American 260-day divinatory calendar, yet the latter was nevertheless integrated, through the Calendar Round, into a larger calendrical system that also included a cycle approximating the actual solar year. That is clearly not the case with the Indonesian week-calendar. While Indonesians use the Hindu lunisolar *saka* year, they have yet to integrate the *odalan* cycle into it. Being based entirely on weekly cycles created by human beings, the Indonesian week-calendar is a rare example of an exclusively artificial time-reckoning system that is totally disregardful of nature and its rhythms. It is essentially a calendar where neither the seasons nor the lunar phases play any role whatsoever. As such, it is probably the most remarkable calendar ever invented, a unique manifestation of the workings of the rational human mind as well as of humans' capability of living in accordance with entirely artificial rhythms which they create. And that, basically, is what the week is all about.

    The Javanese six-day *sadwara*, the Maya thirteen-day divinatory week, the Nigerian four-day market week, and the Baha'i nineteen-day cycle of religious reunions are essentially only different cultural manifestations of a single human theme. In examining these and other weekly cycles,[68] I definitely did not attempt to exhaust human beings' tremendous calendrical creativity and ingenuity. Rather, I tried to demonstrate, through a selection of a few particularly instructive examples, that the range of the phenomenon "week" is much wider than we normally assume.

    The overwhelming cross-cultural variability in the rhythm of life within a single species clearly cannot be attributed no nature and is an obvious result of the human interference with the natural order of things. The range of the phenomenon "week" can be so wide only because it is essentially an institution created by human beings and not part of nature. The main purpose of this chapter was to demonstrate that, despite being so rooted in our everyday life, the seven-day week is not the inevitable fact of nature that it appears to be. Now, after having concluded this brief tour around the world and throughout history, we can definitely see this cycle for what it is—a mere social convention.

CHAPTER FOUR

The Harmonics of Timekeeping

Despite the major role it plays in our life, the week is only one of several components of our timekeeping system. Events such as birthdays and national independence days, for example, are usually observed on fixed annual dates, regardless of the particular day of the week on which they fall. To put it more generally—the temporal location of events in our life is fixed in accordance with a number of different calendrical principles, some of which are totally independent of one another. Thus, for example, "all our daily routines find themselves ruled by two different systems, first, the days of the year, and then the days of the week. The two roll by independently of each other."[1] Fixing events on particular days of the week and on particular annual dates are two rather distinct practices, because they are essentially based on two time-reckoning and dating frameworks that are totally independent of one another, namely the week and the year.[2]

The week evolved quite independently of other building blocks of our timekeeping system. Consequently, despite having coexisted with our calendar month and year for nearly two thousand years, it has yet to be mathematically coordinated with them. For example, with the single exception of the month of February, dividing any calendar month or year into seven-day weeks always leaves some remainder. Our months consist of either two or three days in excess of four weeks, whereas our years always include either one or two days in excess of fifty-two weeks. (Curiously enough, the card deck also consists of fifty-two cards plus a "joker," which may suggest its possible origin as a symbolic representation of a week-calendar, with the fifty-two cards representing the fifty-two complete weeks of the calendar year, and the joker representing the extra day. Note also, incidentally, that the numerical values of the fifty-two cards add up to 364, a number which, together with the joker, equals the total number of days in the calendar year.[3])

To this day, the seven-day week remains a sort of stepchild within an otherwise highly ordered calendrical family, the members of which are all interrelated rather "neatly" as precise multiples of one another. It seems to be the only cycle that somehow manages to "spoil" an otherwise perfect time-measurement and time-reckoning system, wherein each unit basically consists of a complete number of smaller units and its beginning always coincides with the beginnings of all the units below it. Our calendar year, for example, consists of a complete number of months, and its beginning coincides with the beginning of a month, January. That is also true of the relations among the month, the day, the hour, the minute, and the second. Weeks, however, do not have to be completed in order for a new month or year to begin: "Across this ordered system runs that intruder the week, consisting indeed of a fixed number of complete days, but paying no regard to months or years. The moment that begins a new year, begins also a new month, a new day, and a new hour, but only once in five years, at the least, a new week."[4]

The complications resulting from the lack of synchrony between the week and the month are numerous. Consider, for example, how inconvenient it is for employees paid by the month to budget their expenses on a regular weekly basis, when, out of the very same paycheck, they have to make four weekly trips to the supermarket on some months yet five on others. Employees paid by the week, however, do not have an easier time having to pay their monthly telephone, gas, and electric bills sometimes out of five paychecks yet sometimes out of only four.[5] (Given that some months have four weekly paydays while others have five, it is also very difficult for businesses which pay their employees on a weekly basis to calculate their monthly outlay.[6]) Consider, also, some of the problems that arise within the context of scheduling treatment. To take the particular case of tuberculosis patients,

> many patients interpret the "three-month conference" as meaning a twelve-week period, whereas the manner in which the staff schedules

these conferences usually makes them thirteen-week periods. As a result, many patients believe that they have been set back one week on their conference and plead with the doctor to make up for this lost time by scheduling their next conference a week earlier.[7]

Finally, note how annoying we find the oscillation between 28-day and 35-day waiting periods when our kitchen is full of cockroaches and our apartment is regularly fumigated "every third Thursday of the month."

The lack of synchrony between the week and the year presents further inconveniences, one of which is the difference in the week-day/weekend composition of the very same calendar month in two different years. An April that begins on a Thursday, for example, includes twenty-two working days (assuming that Saturday and Sunday are days off), whereas one that begins on a Saturday includes only twenty. This is one of the problems statisticians face when trying to compare any given calendar month across different calendar years.[8] How can a church, for example, compare the attendance figures of two consecutive months of October, when one of them includes five Sundays while the other includes only four?[9] Similarly, within the context of a six-day workweek, when Saturday takings constitute almost one quarter of the entire weekly turnover, how can a business compare the sales figures of two consecutive months of March, when one of them includes five Saturdays and the other one only four?[10]

The Sunday Letter

One obvious result of the lack of synchrony between the week and the year is the fact that annual dates "drift" across the different days of the week, "wandering" every year one or two days away from the particular day on which they fell the year before. Consequently, one must refer to a calendar in order to determine the day of the week on which New Year's Day falls, as well as the precise dates on which Election Day and Thanksgiving Day fall, in any given year. Likewise, given that the American school year is usually tied to Labor Day (which, being fixed on a particular day of the week, falls on different dates in different years), consider also the relatively wide range of annual dates across which a day such as "Friday of the sixth week of school" might drift.[11]

This lack of agreement between days and dates has always been particularly inconvenient for the Church, some of the major festivals on which are fixed on particular days of the week and thus drift across different annual dates in different calendar years. The classic example of such a "movable" festival is Easter, which is celebrated on the Sunday following the full moon which coincides with, or falls next after, the vernal equinox.[12] Some other examples are Ash Wednesday, Ascension Day, and Pentecost, which are observed on the seventh Wednesday before Easter, and on the sixth Thursday and seventh Sunday after it, respectively. It is therefore within the Church that one might expect to find such great interest in devising a method of calculating the relations between particular dates and particular days of the week for different calendar years. And, indeed, it is the Church that deserves the credit for having devised the perpetual almanac of letters which is mainly known for its central feature, the Sunday (or Dominical) Letter.

In their calendars known as *fasti*, the ancient Romans used to designate each day by one of a series of eight letters running from A through H.[13] These letters were repeated successively throughout the calendar year, with January 1 always being designated by the letter A, so that a fixed association between particular annual dates and particular letters could be maintained on a permanent basis. These letters were known as "nundinal letters" since one of them designated all the occurrences of the market day (*nundinae*) within any given calendar year, and there were eight of them since the weekly market day was held every eight days.

Sometime around the middle of the fourth century, this ancient Roman practice of designating the days of the calendar year by a sequence of letters, with January 1 always being designated by the letter A, was also adopted by the Roman Catholic church.[14] The only significant modification introduced, along with the transition from the Roman eight-day market week to the ecclesiastical seven-day week, was that a seven-letter A–G sequence came to replace the original eight-letter A–H sequence. These letters, which can be seen in nearly every premodern calendar, were successively assigned to all the days of the calendar year (with the single exception of February 29), so that the associations between particular annual dates and particular day letters could be fixed on a permanent basis and thus be valid for any given calendar year. As we can see in Figure 5, since January 1 would always be designated by the letter A, February would always begin with the letter D, March with D, April with G, May with B, June with E, July with G, August with C, September with F, October with A, November with D, and December with F. Throughout the ages, a number of Latin as well as English catch verses have evolved as mnemonic devices that help people memorize the letters designating the first day of each month of the calendar

FIGURE 5   The Perpetual Almanac of Day Letters

A	B	C	D	E	F	G
Jan. 1	Jan. 2	Jan. 3	Jan. 4	Jan. 5	Jan. 6	Jan. 7
Jan. 8	Jan. 9	Jan. 10	Jan. 11	Jan. 12	Jan. 13	Jan. 14
Jan. 15	Jan. 16	Jan. 17	Jan. 18	Jan. 19	Jan. 20	Jan. 21
Jan. 22	Jan. 23	Jan. 24	Jan. 25	Jan. 26	Jan. 27	Jan. 28
Jan. 29	Jan. 30	Jan. 31	Feb. 1	Feb. 2	Feb. 3	Feb. 4
Feb. 5	Feb. 6	Feb. 7	Feb. 8	Feb. 9	Feb. 10	Feb. 11
Feb. 12	Feb. 13	Feb. 14	Feb. 15	Feb. 16	Feb. 17	Feb. 18
Feb. 19	Feb. 20	Feb. 21	Feb. 22	Feb. 23	Feb. 24	Feb. 25
Feb. 26	Feb. 27	Feb. 28	Mar. 1	Mar. 2	Mar. 3	Mar. 4
Mar. 5	Mar. 6	Mar. 7	Mar. 8	Mar. 9	Mar. 10	Mar. 11
Mar. 12	Mar. 13	Mar. 14	Mar. 15	Mar. 16	Mar. 17	Mar. 18
Mar. 19	Mar. 20	Mar. 21	Mar. 22	Mar. 23	Mar. 24	Mar. 25
Mar. 26	Mar. 27	Mar. 28	Mar. 29	Mar. 30	Mar. 31	Apr. 1
Apr. 2	Apr. 3	Apr. 4	Apr. 5	Apr. 6	Apr. 7	Apr. 8
Apr. 9	Apr. 10	Apr. 11	Apr. 12	Apr. 13	Apr. 14	Apr. 15
Apr. 16	Apr. 17	Apr. 18	Apr. 19	Apr. 20	Apr. 21	Apr. 22
Apr. 23	Apr. 24	Apr. 25	Apr. 26	Apr. 27	Apr. 28	Apr. 29
Apr. 30	May 1	May 2	May 3	May 4	May 5	May 6
May 7	May 8	May 9	May 10	May 11	May 12	May 13
May 14	May 15	May 16	May 17	May 18	May 19	May 20
May 21	May 22	May 23	May 24	May 25	May 26	May 27
May 28	May 29	May 30	May 31	Jun. 1	Jun. 2	Jun. 3
Jun. 4	Jun. 5	Jun. 6	Jun. 7	Jun. 8	Jun. 9	Jun. 10
Jun. 11	Jun. 12	Jun. 13	Jun. 14	Jun. 15	Jun. 16	Jun. 17
Jun. 18	Jun. 19	Jun. 20	Jun. 21	Jun. 22	Jun. 23	Jun. 24
Jun. 25	Jun. 26	Jun. 27	Jun. 28	Jun. 29	Jun. 30	Jul. 1
Jul. 2	Jul. 3	Jul. 4	Jul. 5	Jul. 6	Jul. 7	Jul. 8
Jul. 9	Jul. 10	Jul. 11	Jul. 12	Jul. 13	Jul. 14	Jul. 15
Jul. 16	Jul. 17	Jul. 18	Jul. 19	Jul. 20	Jul. 21	Jul. 22
Jul. 23	Jul. 24	Jul. 25	Jul. 26	Jul. 27	Jul. 28	Jul. 29
Jul. 30	Jul. 31	Aug. 1	Aug. 2	Aug. 3	Aug. 4	Aug. 5
Aug. 6	Aug. 7	Aug. 8	Aug. 9	Aug. 10	Aug. 11	Aug. 12
Aug. 13	Aug. 14	Aug. 15	Aug. 16	Aug. 17	Aug. 18	Aug. 19
Aug. 20	Aug. 21	Aug. 22	Aug. 23	Aug. 24	Aug. 25	Aug. 26
Aug. 27	Aug. 28	Aug. 29	Aug. 30	Aug. 31	Sep. 1	Sep. 2
Sep. 3	Sep. 4	Sep. 5	Sep. 6	Sep. 7	Sep. 8	Sep. 9
Sep. 10	Sep. 11	Sep. 12	Sep. 13	Sep. 14	Sep. 15	Sep. 16
Sep. 17	Sep. 18	Sep. 19	Sep. 20	Sep. 21	Sep. 22	Sep. 23
Sep. 24	Sep. 25	Sep. 26	Sep. 27	Sep. 28	Sep. 29	Sep. 30
Oct. 1	Oct. 2	Oct. 3	Oct. 4	Oct. 5	Oct. 6	Oct. 7
Oct. 8	Oct. 9	Oct. 10	Oct. 11	Oct. 12	Oct. 13	Oct. 14
Oct. 15	Oct. 16	Oct. 17	Oct. 18	Oct. 19	Oct. 20	Oct. 21
Oct. 22	Oct. 23	Oct. 24	Oct. 25	Oct. 26	Oct. 27	Oct. 28
Oct. 29	Oct. 30	Oct. 31	Nov. 1	Nov. 2	Nov. 3	Nov. 4
Nov. 5	Nov. 6	Nov. 7	Nov. 8	Nov. 9	Nov. 10	Nov. 11
Nov. 12	Nov. 13	Nov. 14	Nov. 15	Nov. 16	Nov. 17	Nov. 18
Nov. 19	Nov. 20	Nov. 21	Nov. 22	Nov. 23	Nov. 24	Nov. 25
Nov. 26	Nov. 27	Nov. 28	Nov. 29	Nov. 30	Dec. 1	Dec. 2
Dec. 3	Dec. 4	Dec. 5	Dec. 6	Dec. 7	Dec. 8	Dec. 9
Dec. 10	Dec. 11	Dec. 12	Dec. 13	Dec. 14	Dec. 15	Dec. 16
Dec. 17	Dec. 18	Dec. 19	Dec. 20	Dec. 21	Dec. 22	Dec. 23
Dec. 24	Dec. 25	Dec. 26	Dec. 27	Dec. 28	Dec. 29	Dec. 30
Dec. 31						

year. They are essentially chains of twelve words, the initials of which correspond to the letters designating the first days of the twelve months:

At Dover Dwells George Brown Esquire,
Good Christopher Finch, And David Fryar.

Alta Domat Dominus, Gratis Beat Equa
Gerentes Contemnit Fictos, Augebit Dona Fideli.

Astra Dabit Dominus Gratisque Beabit Egenos;
Gratia Christicolae Feret Aurea Dona Fideli.[15]

Dividing a 365-day calendar year into 7-day weeks always leaves a 1-day remainder in excess of the fifty-two complete weeks. Consequently, since each calendar year is supposed to begin with the letter A, the association between particular letters and particular days of the week necessarily changes from one year to the next. The Romans, incidentally, had to face the very same problem, since dividing a 365-day calendar year into 8-day market cycles also leaves a remainder of 5 days in excess of the forty-five complete market weeks. Since the *nundinae* was held on a regular weekly basis every eight days, it was obviously designated by a different nundinal letter every year. Like the Roman market week, the ecclesiastical week is also entirely independent of the calendar year, and Sunday, like the market day, also recurs on a regular weekly basis—every seven days, with no exception whatsoever. As a result of that, each ecclesiastical calendar year is also characterized by a different letter designation of Sunday.

The particular Sunday-Letter designation of any given year constitutes the necessary bridge between the abstract almanac of day letters and actual historical time. If I do not know the particular Sunday-Letter designation of a given year, the perpetual almanac presented in Figure 5 is of no use to me whatsoever. Once I know it, however, calculating the relation between any particular date and day of the week within that year becomes a very easy task. On the column right below the Sunday Letter, that almanac provides me with all the dates that fall on Sunday that particular year. From there, I can easily proceed to identify all the Mondays, Tuesdays, and so forth. Thus, for example, if I wish to find out the precise date on which Labor Day falls in 1983, a year which I know to be designated by the Dominical Letter B, I can easily figure out that all the Mondays during that year would be located in column C. Then, down that column, I find out that the first Monday in September (that is, Labor Day) falls on September 5. Along similar lines, in order to find out the day of the week on which October 28 falls that year, I first locate that date in column G, and then figure out that it would be a Friday.

The reason that the year 1983 was designated by the Sunday Letter B is that the first Sunday in it fell on January 2. Along similar lines, the years 1982 and 1981 were designated by the letters C and D because the first Sundays in them fell on January 3 and January 4, respectively. Since the 365-day calendar year includes one day in excess of fifty-two weeks, successive years generally begin on successive days of the week. The year 1981, for example, began on a Thursday, the year 1982 on a Friday, and the year 1983 on a Saturday.

As a result of all that, Sunday Letters normally flow in a continuous retrograde fashion—from E to D, from A to G, and so on. This pattern of "regression" of Dominical Letters is a permanent one, and, throughout the ages, has been captured in a number of Latin catch verses, composed by eminent scholars such as the Venerable Bede, Johannes de Sacrobosco, and Julius Scaliger. These verses usually consist of chains of twenty-eight words, the initials of which correspond to a cycle of twenty-eight Dominical Letters designating successive years:

> Fallitur Eva Dolo, Cibus *Adae Gaudia Finit
> Et *Cum Botrus Adhuc Germinet, *Eua Dolet
> Christus Bella *Gerit, Finitur Eo Duce *Bellum
> Ad Grauidam Fit *Dux, Cuncta Beavit Ave.
>
> Fert Ea Dux Cor *Amat Gens Frons E *Coluit Bis
> Ars Genus *Est De Cordo Bono *Gignit Ferus Ensis
> Dicta *Beant Aqua Gens Fons *Dat Cunctis Bonus Auctor. [16]

The asterisks in the above verses are designed to indicate that the cycle of day letters is regularly interrupted, once every four years, by certain "leaps"—from C to A, from B to G, and so on. In order to understand those leaps (which are also responsible for the concept "leap year"[17]), we ought to remember that, despite having been systematically ignored by the Church in its almanac of day letters, never having been designated by any letter (see Figure 5), an extra, 366th day (February 29) is actually added to the calendar every four years. Between normal calendar years, which involve only one day in excess of fifty-two complete seven-day weeks, Dominical Letters recede at a pace of one letter a year, since successive years always begin on successive days of the week. Leap years, however, include two days in excess of fifty-two weeks. Therefore, between leap years and normal calendar years, Sunday Letters recede two letters instead of one. As a result, leap years are always designated by two successive Sunday Letters.

As can be seen from the above catch verses, as well as in Figure 6, which anchors the cycle of Sunday Letters within actual historical time, it is indeed once every four years that the continuous one-letter-a-year pace of the retrograde flow of the seven letters is regularly interrupted by double-letter leap years. As a result, it takes the cycle of day letters twenty-eight (that is, 7 × 4) years to complete one round. The length of this twenty-eight-year cycle of Dominical Letters (generally known as the *Solar Cycle*) is most conspicuously manifested in the number of words constituting the above catch verses as well as in the number of rows constituting the table in Figure 6.

Note, however, that, ever since the Gregorian reform of the calendar in 1582, not all calendar years that are precise multiples of 4 have an extra, 366th day. From among century years, only those that are divisible by 400 are considered leap years. (Thus, whereas the years 1600 and 2000 are intercalated, the years 1700, 1800, and 1900 are not.) The fact that only one out of every four century years is intercalated obviously interrupts the continuous flow of the Solar Cycle of Dominical Letters. As a result, the table in Figure 6 is organized in four separate columns for the four different types of centuries within each 400-year cycle, and also includes an extra, twenty-ninth row at the top for century years.

Finally, using both Figures 5 and 6, I can find out the particular day of the week that corresponds to any date within any given calendar year. Suppose, for example, I wish to find out the exact day on which October 27, 1970 fell. Using Figure 6, I first locate the twentieth century on the second column from the right on the left half of the table. Then, on the right half of that table, I identify the year ". . 70" on the fifteenth row from the top. Correlating the appropriate row and column, I find out that the year 1970 was designated by the Dominical Letter D. Turning then to Figure 5, I know that column D contains all the Sundays of that calendar year. I then locate the date October 27 down column F, and easily figure out that October 27, 1970 fell on a Tuesday. (Leap years are designated by two Sunday Letters. The first letter applies to the months of January and February whereas the second applies to the rest of the year.)

Under the present conditions, short of having to refer to an actual calendar, this rather cumbersome routine is still the fastest noncomputerized method of identifying the days of the week that correspond to specific annual dates. It seems as if this inconvenience is a price we must pay for having invented the seven-day week. And yet, is the lack of synchrony between the week and the longer calendrical cycles indeed inevitable?

Curiously enough, this lack of synchrony resembles that which exists between various cycles of musical intervals. Let us consider again the "calendrical leftovers" mentioned earlier, namely the 1.53-day remainder of the lunar month in excess of four weeks and the

FIGURE 6   The Sunday-Letter Designation of Calendar Years

CENTURIES				YEARS WITHIN EACH CENTURY			
1700 2100	1800 2200	1500 1900 2300	1600 2000 2400				
C	E	G	B A	. . 00			
B	D	F	G	. . 01	. . 29	. . 57	. . 85
A	C	E	F	. . 02	. . 30	. . 58	. . 86
G	B	D	E	. . 03	. . 31	. . 59	. . 87
F E	A G	C B	D C	. . 04	. . 32	. . 60	. . 88
D	F	A	B	. . 05	. . 33	. . 61	. . 89
C	E	G	A	. . 06	. . 34	. . 62	. . 90
B	D	F	G	. . 07	. . 35	. . 63	. . 91
A G	C B	E D	F E	. . 08	. . 36	. . 64	. . 92
F	A	C	D	. . 09	. . 37	. . 65	. . 93
E	G	B	C	. . 10	. . 38	. . 66	. . 94
D	F	A	B	. . 11	. . 39	. . 67	. . 95
C B	E D	G F	A G	. . 12	. . 40	. . 68	. . 96
A	C	E	F	. . 13	. . 41	. . 69	. . 97
G	B	D	E	. . 14	. . 42	. . 70	. . 98
F	A	C	D	. . 15	. . 43	. . 71	. . 99
E D	G F	B A	C B	. . 16	. . 44	. . 72	
C	E	G	A	. . 17	. . 45	. . 73	
B	D	F	G	. . 18	. . 46	. . 74	
A	C	E	F	. . 19	. . 47	. . 75	
G F	B A	D C	E D	. . 20	. . 48	. . 76	
E	G	B	C	. . 21	. . 49	. . 77	
D	F	A	B	. . 22	. . 50	. . 78	
C	E	G	A	. . 23	. . 51	. . 79	
B A	D C	F E	G F	. . 24	. . 52	. . 80	
G	B	D	E	. . 25	. . 53	. . 81	
F	A	C	D	. . 26	. . 54	. . 82	
E	G	B	C	. . 27	. . 55	. . 83	
D C	F E	A G	B A	. . 28	. . 56	. . 84	

1.2422-day remainder of the solar year in excess of fifty-two weeks. Are they essentially so different from such infinitesimal, yet quite bothersome and mathematically puzzling, musical intervals as the Pythagorean comma (the excess of twelve fifths over seven octaves), the schisma (the excess of eight perfect fifths and a true major third over five octaves), or the diesis (the excess of an octave over the sum of three major thirds)?[18]

Throughout history, the process of "rationalizing" music has largely consisted of various attempts to incorporate those "irrational" intervals into a perfectly symmetrical system of tonal physics.[19] Interestingly enough, the very same "rational" mentality and spirit has also manifested itself within the domain of calendrics. Throughout history, there have been numerous parallel attempts to synchronize the week with the longer calendrical cycles and thus fix the relations between particular dates and particular days of the week on a permanent, "rational" basis.

The Perpetual 364-Day Calendar

The simplest way of synchronizing the weekly and annual cycles is, obviously, the establishment of a calendar year that is a precise multiple of the week, just like the Central American and Indonesian 260-day and 210-day calendar "years." In the particular case of the seven-day week, the closest such approximation of the actual 365.2422-day solar cycle would be a 364-day calendar year which consists of precisely fifty-two weeks with no remainder whatsoever.

Two Jewish pseudoepigraphic sources dating from the second century B.C., *The Book of Enoch* and *The Book of Jubilees*, seem to indicate that such a calendrical solution was actually tried out at least two thousand years ago in ancient Judea:

> And the sun and the stars bring in all the years exactly, so that they do not advance or delay their position by a single day unto eternity; but complete the years with perfect justice in 364 days.[20]

> And all the days of the commandment will be *two and fifty weeks* of days, and these will make the entire year complete. . . . And command thou the children of Israel that they observe the years according to this reckoning—three hundred and sixty-four days, and these will constitute a complete year.[21]

Both sources describe a 364-day calendar year that was subdivided into four 91-day trimesters. At times, the year is described as consisting of twelve 30-day months with four extra days interspersed between trimesters.[22] At other times, trimesters are depicted as consisting of two 30-day months and one 31-day month.[23] Most crucially significant, however, is the fact that the calendar year was subdivided into four 13-week blocks.[24] Revolving around an annual cycle that consisted of precisely fifty-two weeks (as well as of four 13-week trimesters), this calendar—just like the Baha'i, Central American, and Indonesian week-calendars—was obviously based on the week as its most fundamental building block.

With the calendar year being defined as a precise multiple of the week, years could always begin on the same day of the week and the association between particular annual dates and particular days of the week could be fixed on a permanent basis. All annual festivals, for example, could thus be regularly associated not only with a particular date, but also with a particular day of the week. In short, this 364-day calendar was essentially a *perpetual calendar* that would apply to any given calendar year without ever having to be altered. In fact, as we can see from Figure 7, since calendar years were subdivided into precisely four 13-week trimesters which consisted of a single, uniform pattern of two 30-day months followed by a 31-day month, it was actually a perpetual 91-day calendar that would essentially apply to any given trimester in any given year. In order to guarantee that this *perfect synchrony between the week and the year* would never be interrupted, this calendar was most probably synchronized with the actual solar 365.2422-day cycle through the establishment of intercalary intervals that were themselves precise multiples of the week. These were most likely thirty-five-day periods that were inserted at the end of every twenty-eight years.[25]

As seen in Figure 7, the 364-day calendar year most probably began on Wednesday, a most peculiar fact, given that the Jewish week begins on Sunday and that Wednesday's traditional Hebrew name has always indicated that it is only the fourth day. The explana-

FIGURE 7   The Perpetual 364-Day Calendar

Months

	I, IV, VII, X					II, V, VIII, XI				III, VI, IX, XII			
Days														
Wednesday	1	8	15	22	29	6	13	20	27		4	11	18	25
Thursday	2	9	16	23	30	7	14	21	28		5	12	19	26
Friday	3	10	17	24		1	8	15	22	29	6	13	20	27
Saturday	4	11	18	25		2	9	16	23	30	7	14	21	28
Sunday	5	12	19	26		3	10	17	24	1	8	15	22	29
Monday	6	13	20	27		4	11	18	25	2	9	16	23	30
Tuesday	7	14	21	28		5	12	19	26	3	10	17	24	31

tion for this oddity may be found in the biblical account of the fourth day of the creation of the world: "And God said: 'Let there be lights in the firmament of the heaven to divide the day from the night; and let them be for signs, and for seasons, and for days and years.' "[26] Given this traditional Jewish emphasis on the calendrical function of the luminaries, it is quite understandable that a calendar year might begin on the day of the week on which they were believed to have been created.[27]

Interestingly enough, in the passage that parallels the above biblical account of the fourth day of the Creation, the author of *Jubilees* makes it quite clear that, of the two major luminaries, it was definitely the sun, and not the moon, that was designated to regulate the calendar: "And god appointed the sun to be a great sign on the earth for days and for sabbaths and for months and for feasts and for years and for sabbaths of years and for jubilees and for all seasons of the years."[28] Note also, in this regard, that the third section of *Enoch*, titled "The Book of Heavenly Luminaries," opens with a chapter on the sun and only then proceeds to discuss the moon.[29] Also, particularly given the association of the legendary antediluvian figure Enoch with the establishment of the calendar,[30] it is probably not a coincidence that his life on earth is traditionally recorded as having lasted 365 years.[31] The symbolic association between the number 365 and an essentially solar calendar is quite obvious.

And, indeed, it is only within the context of the contrast between solar and lunar (or lunisolar) calendars that the full social significance of the perpetual 364-day calendar can be appreciated. As the following passage from *Jubilees* seems to indicate, that contrast had considerable political overtones:

> And command thou the children of Israel that they observe the years according to this reckoning—three hundred and sixty-four days, and these will constitute a complete year, and they will not disturb its time from its days and from its feasts; for everything will fall out in them according to their testimony, and they will not leave out any day nor disturb any feasts. But if they do neglect and do not observe them according to His commandment, then they will disturb all the seasons, and the years will be dislodged from this order, and they will disturb the seasons and the years will be dislodged and they will neglect their ordinances. And all the children of Israel will forget, and will not find the path of the years, and will forget the new moons, and seasons, and sabbaths, and they will go wrong as to all the order of the years. . . lest they forget the feasts of the covenant and *walk according to the feasts of the Gentiles* after their error and after their ignorance. For there will be those who will assuredly make observations of *the moon— how it disturbs the seasons and comes in from year to year ten days too soon.* For this reason the years will come upon them when they

will disturb the order, and make an abominable day the day of testimony, and an unclean day a feast day, and they will confound all the days, the holy with the unclean, and the unclean day with the holy; for they will go wrong as to the months and sabbaths and feasts and jubilees. For this reason I command and testify to thee that thou mayst testify to them; for after thy death thy children will disturb them, so that *they will not make the year three hundred and sixty-four days only, and for this reason they will go wrong as to the new moons and seasons and sabbaths and festivals, and they will eat all kinds of blood with all kinds of flesh.*[32]

At the very end of this passage, the author of *Jubilees* makes an explicit connection between the abandonment of the 364-day calendar—an act he earlier associates with "walking according to the feasts of the Gentiles after their error and after their ignorance"—and the total rejection of the very essence of Judaism. Throughout the passage, he keeps stressing the broader symbolic and cultural significance of what appear, on the surface, to be merely calendrical choices.

It is quite possible that Jews had originally used the solar, 364-day calendar long before they were introduced to the lunar calendar,[33] and it was as a distinctively Jewish institution that the foremost Jewish fundamentalists, the Sadducees, were probably trying to preserve that calendar. However, by the second century B.C, when both *Enoch* and *Jubilees* were most probably written, the 364-day calendar could only be struggling for its survival. It was during that period that the Hellenistic lunisolar, 354-day calendar was gaining considerable popularity in Judea, particularly among the Sadducees' main rivals, the relatively modernistic Pharisees, who were somewhat more receptive to the Hellenization of traditional Jewish cultural institutions such as the calendar. Both *Enoch* and *Jubilees* were most probably written by Sadducees, whose main target, when condemning those who followed non-Jewish calendrical practices, were Pharisees. The vigor with which the author of *Jubilees* defends the 364-day calendar while defying the increasingly popular lunisolar 354-day calendar only serves to highlight the considerable political undercurrents of the calendrical conflict between those two rivaling parties in second-century B.C. Judea.

The calendrical conflict between the Sadducees and the Pharisees revolved around the problem of when to start counting the days leading toward the celebration of the Feast of Weeks,[34] and its relevance to our discussion becomes clearer once we realize the major role played by the week in the Saducean solution to this problem. (The culmination of *The Book of Jubilees* in the introduction of the Sabbath[35] also serves to highlight the centrality of the week to the

Sadducees.) The Pharisees used to start counting those days from the day immediately following Passover Eve, which always coincides with the full moon, regardless of the particular day of the week on which it happens to fall. By contrast, the Sadducees would start counting them only from the Sunday immediately following that festival. (It is quite possible that this Sadducean custom inspired the Church's original practice of observing Easter on the Sunday immediately following Passover Eve.[36]) This Sadducean practice obviously reflects a strong desire to dissociate Jewish festivals from the lunar cycle and anchor their observance permanently on particular days of the week. It would obviously be enhanced by the establishment of a perpetual 364-day calendar, which would allow for a perfect synchrony between the week and the year and, thus, also between particular dates and particular days of the week.

The Sadducees lost that battle of the calendars and, to this day, all Jewish festivals are fixed in accordance with the lunisolar calendar. However, as late as the first century of the present era, the perpetual 364-day calendar was still being used by the monastic community generally known as the Dead Sea Sect.[37] According to medieval historians, these monks always observed both New Year's Day and Passover on Wednesday. Not only does this correspond perfectly to the calendar presented in Figure 7 (the above festivals being the first and fifteenth days of months I and VII); it is also only in a perpetual calendar allowing for a perfect synchrony between the week and the year that particular annual dates could always fall on the same day of the week. Moreover, in the Qumran caves where this sect dwelled in the Judean desert, fragments of *The Book of Jubilees* have also been found, along with the famous so-called Dead Sea Scrolls. The latter mention fifty-two "fathers of the congregation," their number being strikingly suggestive of the number of weeks making up the solar 364-day calendar year. (By contrast, the lunisolar 354-day calendar year includes only less than fifty-one weeks.) The scrolls also mention twenty-six "chiefs of courses," and it appears that this fundamentalist monastic community divided the liturgical calendar year into twenty-six courses (*mishmarot*) in accordance with which priestly families would rotate service setting out the incense and bringing in the sacrifices. This peculiar arrangement contrasted with the traditional schedule of the Temple in Jerusalem, which involved only twenty-four priestly courses every year, as priests would normally officiate for half a month at a time. It seems to indicate that the Qumran monks definitely reckoned liturgical time in terms of weeks, regarding their calendar year as being made up of twenty-six double-weeks, rather than of twenty-four half-months, as it would be regarded in Jerusalem.

Along with the Qumran community, the perpetual 364-day calendar disappeared from Judea some 1,900 years ago, and, with the single significant exception of the Ethiopian calendar,[38] it has been all but forgotten. And yet, the conception of the year as a fifty-two-week cycle is still preserved to this day in business accounting systems that are based on thirteen-week trimesters,[39] as well as in the organization of television programming along four 13-week "seasons." It has also manifested itself in one other significant form. There, however, the synchrony between the week and the year (and, thus, between days and dates) has been accomplished through treating an actual 365¼-day year as if it were a 364-day cycle.

"Blank" Days

The synchrony between the annual and weekly cycles can obviously be accomplished through counting one particular day of the week repeatedly for several days in a row once a year. Thus, for example, in order to eliminate the 2-day and 3-day remainders that are left after dividing the 210-day *odalan* cycle into the 4-day, 8-day, and 9-day weeks, Indonesians triplicate the days Jaya (from the four-day week) and Kala (from the eight-day week) and quadruplicate the day Danggú (from the nine-day week) once every "year." (See the annual interruption of the continuous flow of those three weekly cycles in the *ukus* Dungulan and Sinta in Figure 4.) However, one could also accomplish the synchrony between the week and the year through the application of a diametrically opposite strategy, namely regarding those days of the calendar year that are in excess of the highest precise multiple of the week as *blank days* that would not be reckoned as parts of any particular week. There is a traditional Chinese story about a dying man who left his three sons eleven sheep, stipulating that one half should go to the oldest, one quarter to the middle, and two thirds of what is left to the youngest. The sons proceeded to borrow a twelfth sheep from a neighbor, took six, three, and two sheep each, and then returned the extra sheep to its owner.[40] In quite an analogous fashion, one could also remove a few days from the calendar year, arrange the days that are left in complete weekly cycles, and then bring back the extra days as a supplementary add-on. As those "blank" days would not be reckoned as parts of any particular week, the calendar year would be regarded as if it consisted indeed of a precise number of complete weeks with no remainder.

Such a calendrical arrangement is exemplified by the subdivision of the Baha'i 365-day calendar year into nineteen 19-day weeks with

4 extra intercalary days (the *ayyám-i-há*) and of the Central American solar 365-day calendar year into eighteen 20-day weeks with 5 extra intercalary days. Given the centrality of the week to these two week-calendars, those remaining days that do not belong to any particular week obviously constitute a residual category. Not surprisingly, these days—collectively known as "days without names" (*xma kaba kin*) among the Maya and as "hollow" or "superfluous" days (*nemontemi*) among the Aztecs—were associated with the demon of evil and observed as days of abstinence. Anyone who was born on those five sinister days was destined to a miserable life.[41]

The 1.2422-day remainder in excess of the fifty-two complete 7-day weeks of our own 365¼-day calendar year can be handled in a rather similar manner. That would involve excluding 1 day of the year (or 2 days, in leap years) from the reckoning of weeks, and regarding the year as if it were indeed a 364-day cycle consisting of precisely fifty-two weeks with no remainder. This idea, seriously considered in 1834 by the Italian priest Marco Mastrofini, was the highlight of a calendar-reform competition sponsored in 1887 by the French Astronomical Society. The first two prizes were won by two French citizens, Gaston Armelin and Emil Hanin, whose proposed calendars were essentially based on four identical thirteen-week trimesters plus an additional "blank" New Year's Day (as well as a Leap Day, in leap years), to be inserted between December and January.[42] Four very similar calendars were proposed, between 1900 and 1912, by L. A. Grosclaude and Arnold Kempe from Switzerland, Alexander Philip from England, and Gabriel Nehapetian from Italy.[43] Yet the most famous calendar-reform proposal in the above spirit was undoubtedly the so-called World Calendar, introduced in 1930 by Elisabeth Achelis from the United States and promoted for twenty-five years through the official organ of her World Calendar Association, *The Journal of Calendar Reform*. This calendar was essentially based on the Calendar Reform Bill presented before the British House of Commons shortly before World War I.[44] It generally consisted of four identical thirteen-week trimesters, each one beginning on Sunday and made up of one 31-day month followed by two 30-day months. An additional "blank" Year-End Day (later renamed Worldsday) was inserted between December and January, and a Leap-Year Day was added in leap years between June and July (see Figure 8).

All the above proposals essentially involved a conception of the calendar year as a 364-day cycle consisting of fifty-two complete weeks plus an additional "blank" day that was inserted between December and January and was not reckoned as part of any particular week. (The only differences among them were whether years and trimesters would begin on Sunday or on Monday, whether Leap-

FIGURE 8   The Perpetual Trimestral World Calendar

January April July October							February May August November							March June September December						
S	M	T	W	T	F	S	S	M	T	W	T	F	S	S	M	T	W	T	F	S
1	2	3	4	5	6	7				1	2	3	4						1	2
8	9	10	11	12	13	14	5	6	7	8	9	10	11	3	4	5	6	7	8	9
15	16	17	18	19	20	21	12	13	14	15	16	17	18	10	11	12	13	14	15	16
22	23	24	25	26	27	28	19	20	21	22	23	24	25	17	18	19	20	21	22	23
29	30	31					26	27	28	29	30			24	25	26	27	28	29	30*

* December 30 to be followed by "blank" Worldsday. June 30 to be followed by "blank" Leap-Year Day in leap years.

Year Day would be observed at the end of June or December, and whether the internal monthly pattern of the four identical trimesters would be 31–30–30, 30–30–31, or 35–28–28.) The exclusion of the 365th day of the year (as well as of the 366th one, in leap years) from the reckoning of weeks would allow users of any of these calendars all the cognitive advantages of the perpetual 364-day calendar. Years would always begin on the same day of the week, and the association of particular annual dates with particular days of the week would be fixed on a permanent basis. In the World Calendar, for example, Labor Day would be permanently fixed on September 4, whereas Christmas Day would always fall on Monday (see Figure 8). A temporal formulation such as "Tuesday, October 10" would be unnecessarily reduntant, since October 10 would always fall on Tuesday. Under these circumstances, "reference to a calendar will be entirely unnecessary. Calendar printing will be a formality. One may expect to see 'perpetual calendars,' chiefly for ornamental purposes, engraved in precious metals."[45]

All the above calendars also included the intermediary unit we have already examined in the perpetual 364-day calendar, namely, the 91-day trimester. As the days in excess of four such trimesters were excluded from the annual reckoning of weeks, and as each trimester consisted of precisely thirteen weeks, users of these calendars—just like those of the perpetual 364-day calendar—would only need to memorize the internal weekly composition of a single, generic trimester, consisting of three generic types of months (see Figure 8).

This trimestral harmony, however, did not necessarily entail monthly harmony as well, since the above trimesters consisted of months that were not precise multiples of the week. (The exceptional case of Arnold Kempe's trimesters, which consisted of one 35-day and two 28-day months, also involved more than a single generic type of month.) The permanent association between particular days of the week and particular dates would thus apply only at the level of each one of the three generic types of months within trimesters (first, middle, and last), but not at the level of months in general. In the World Calendar, for example, April 19 would always fall on Thursday, as would January 19, July 19, and October 19 (see Figure 8). May 19, however, would always fall on Sunday, and September 19 on Tuesday.

However, on a number of occasions throughout history, all involving "blank" days that were not reckoned as part of any particular week, the week has been successfully synchronized not only with the year, but with the month as well. The precedent was set by the ancient Egyptians, who included in their 365-day calendar year 5 "epagomenal" days that were not counted as part of any particular 10-day week (or 30-day month, for that matter).[46] Obviously modeled after these days were the five (or six, in leap years) *sansculottides* that were incorporated into the French Republican calendar in 1793.[47] Similarly excluded from the normal annual reckoning of both weeks and months were the five annual holidays introduced by the Soviet government in 1929 along with the abolition of the seven-day week.[48] (The latter, however, were interspersed throughout the calendar year, unlike the Central American, Baha'i, Egyptian, and French Republican "blank"days, all of which were grouped together in one block at the end of the year.) All of the above cases involved the establishment of uniform 30-day months that were precise multiples of the weekly cycle (the Egyptian and French 10-day week and the Soviet 5-day and 6-day weeks), as well as the exclusion of all days in excess of 360 days (the longest precise multiple of both the week and the month) from the annual reckoning of both weeks and months.

The obvious result was a perfect *perpetual synchrony between the week, the month, and the year,* which allowed for a permanent association between particular dates and particular days of the week. Given the introduction of "blank" days, this association applied to annual as well as monthly dates, which actually made a *perpetual monthly calendar* possible (see Figure 9). In the French Republican calendar, for example, a temporal formulation such as "Septidi, Nivôse 27" would be unnecessarily redundant, since the 27th day of any given month would always fall on Septidi. Users of such a calen-

FIGURE 9   The Perpetual Monthly French Republican Calendar

Primidi	Duodi	Tridi	Quartidi	Quintidi	Sextidi	Septidi	Octidi	Nonidi	Décadi
1	2	3	4	5	6	7	8	9	10
11	12	13	14	15	16	17	18	19	20
21	22	23	24	25	26	27	28	29	30

dar only needed to memorize the association between particular dates and particular days within the generic month presented in Figure 9. (In the particular case of the French Republican calendar, the conversion of dates and days into one another was even further facilitated by the fact that the names of the days of the week, which were a reflection of their temporal location within it, also matched the last digit of the corresponding monthly dates. Consider, for example, the rather conspicuous affinity between the numbers 5, 15, and 25 and the name Quintidi, which was associated with the only day of the week on which either the fifth, fifteenth, or twenty-fifth day of any given month could ever fall.) The tremendous cognitive advantages of this simple calendrical arrangement (which obviously did not escape the attention of the architects of the French Republican calendar[49]) can hardly be overexaggerated, considering the extent to which it facilitated the process of juggling two traditionally incompatible levels of temporal reference such as the month and the week.

The synchronization of weeks, months, and years is by no means restricted to calendars involving thirty-day months and ten-day, five-day or six-day weeks. Removing 1 day (or 2, in leap years) from the 365-day calendar year may help us divide the latter not only into uniform trimesters, but also into uniform months, which are precise multiples of the 7-day week. After all, there is no reason why the very same three factors (4, 13, and 7) that have been used for subdividing the 364-day annual cycle into four trimesters which consist of thirteen 7-day weeks could not also be employed for subdividing it into thirteen months which consist of four 7-day weeks (or seven 4-day weeks, as in the case of the Bambala of Central Africa[50]). Admittedly, unlike a twelve-month year, a thirteen-month year cannot be divided into four symmetrical trimesters. At the same time, dividing a 364-day annual cycle into precisely thirteen 28-day months, which can in turn be subdivided into precisely four 7-day weeks, allows for a perfect perpetual synchrony between the week, the month, and the year.

A thirteen-month calendar may have been tried in antiquity[51] and was also part of a calendar-reform proposal made in 1745 by "Hirossa-ap-Iccum" from Maryland.[52] Its most prominent modern manifestations, however, have been the Positivist calendar established in 1849 by the French visionary Auguste Comte[53] and the International Fixed Calendar proposed in 1922 by Moses B. Cotsworth from England.[54] All these calendars involved the exclusion of one annual "blank" day, along with an additional 366th day in leap years, from the reckoning of both weeks and months—Christmas Day and Leap Day in the calendar proposed by "Hirossa-ap-Iccum," the Festival of All the Dead and the Festival of Holy Women in the Positivist calendar, and New Year's Day and Sol Day in the International Fixed Calendar. The remaining 364 days were then divided into precisely thirteen 28-day months, each of which was in turn subdivided into precisely four 7-day weeks, with no remainder being left in either case.

With the 364-day annual cycle as well as the 28-day month being precise multiples of the 7-day week, all weeks, months, and years could be in perfect perpetual synchrony with one another. They would all begin on the same day of the week (Monday in Comte's calendar, Sunday in Cotsworth's), and particular dates would be permanently associated with particular days of the week. This association applied to annual as well as to monthly dates, so that a perpetual calendar of a single generic month (such as the one presented in Figure 10) could again suffice for establishing the relations between dates and days within any given month or year. Thus, for example, in the International Fixed Calendar, the 13th day of any given month would always fall on Friday. (As one might expect of such a "rational" system, it was totally oblivious or insensitive to any irrational superstitions.) The concordance of days and dates (and, along with it, the harmonization of two modes of time reckoning that had traditionally been entirely independent of one another) was again a top item on the agenda of the architects of the above calendars. To quote Meredith Stiles, one of the staunchest supporters of the International Fixed Calendar, "for the first time in history this ancient institution, the seven-day week, would become truly a part of the calendar instead of the independent cycle which it really is at present."[55]

Many of the calendars here examined were seriously considered for use at both national and international levels. Since 1908, numerous reform proposals were presented before the British House of Commons, the U.S. Congress, the International Congress of Chambers of Commerce, and even the League of Nations.[56] And yet, none of these calendars has ever been adopted on a large scale, a fact that is quite striking, particularly given the impressive lists of supporters of some of them. The National Committee of Calendar Simplification for the United States, which promoted the International Fixed Calendar, was founded by the famous American inventor and industrialist George Eastman, and included such prestigious and influential mem-

FIGURE 10   The Perpetual Monthly International Fixed Calendar

Sunday	Monday	Tuesday	Wednesday	Thursday	Friday	Saturday
1	2	3	4	5	6	7
8	9	10	11	12	13	14
15	16	17	18	19	20	21
22	23	24	25	26	27	28

bers as Henry Ford, the Secretary of Labor, the publisher of the *New York Times,* the chief of the United States Weather Bureau, the directors of the Bureau of Standards and the *Nautical Almanac,* and the presidents of Yale University, Cornell University, the Massachusetts Institute of Technology, General Motors, General Electric, the National Geographic Society, the American Museum of Natural History, and the American Bar Association.[57] (In addition to that, an operations calendar based on thirteen 4-week "months" was adopted by several hundred large concerns, including Sears and Roebuck and the Hearst publications.[58]) The World Calendar, at the same time, was officially endorsed by numerous scholarly societies (for example, the American Association for the Advancement of Science, the American Academy of Arts and Sciences, and the American astronomical, mathematical, philosophical, and psychological associations), many presidents of colleges, various commercial organizations (such as the British Chamber of Commerce, the American Industrial Bankers Association, and the American Institute of Accountants), and various religious groups (such as the American Lutheran Church and the General Convention of the Protestant Episcopal Church).[59]

The failure of these calendars to gain official acceptance despite all this support can be explained only by a very deep societal resistance, which was explicitly articulated only by extreme Sabbatarians, Jews as well as Christians, such as Seventh-Day Adventists, the League for Safeguarding the Fixity of the Sabbath, and the Lord's Day Observance Society.[60] Interestingly enough, these groups did not seem at all bothered by the prospect of a perpetual calendar that would essentially revolve around a 364-day cycle that would allow for a perfect perpetual synchrony between the week, the month, and the year. In fact, some of the most vocal opponents of the International Fixed Calendar even offered their own solutions as to how such a cycle might be synchronized with the actual solar 365.2422-day year. However, they seemed to favor the method of intercalation that we examined earlier with respect to the perpetual 364-day calendar, namely, accumulating the 1.2422-day annual surplus in the form of intercalary intervals that would themselves be precise multiples of the week (for example, a 28-day "leap month" that would be inserted at the end of every twenty-three years).[61] It was obviously only the prospect of the interruption of the continuous flow of the week by the introduction of "blank" days that Sabbatarians found objectionable about the World Calendar and the International Fixed Calendar. That was why they regarded the national and international debates about these calendars as actual battles over basic religious freedom. (In fact, the Chief Rabbi of the British Empire even went as far as to publish a full book-length report about the defeat of the Cotsworth–Eastman reform proposal at the League of Nations, which he hailed as nothing less than "a great victory in a fight for liberty second in importance to no other in many a century."[62])

In order to appreciate why Sabbatarians regarded the debates about the World Calendar and the International Fixed Calendar as actual battles over basic religious freedom, we must realize that, if any of those calendars were to be put into effect, the Sabbath and the Lord's Day would no longer be permanently fixed on Saturday and on Sunday, as they had been for thousands of years. With one "blank" day (or two, in leap years) being excluded from the annual reckoning of weeks, the Sabbath, for example, would necessarily drift back one or two days every year, essentially becoming a "floating," "migrating" "nomad" or "wanderer"[63] that might fall on just any day of the week. That, of course, would have been a preposterous idea for any traditionalist, for whom the Sabbath and the Lord's Day had always been synonymous with Saturday and Sunday. Far more critical, however, would have been the practical exclusion of Sabbatarians from active participation in many social affairs in all those years when their holy day would fall on one of the five regular weekdays. Thus, for example, in all those years, observant Jews who would close their businesses during the Sabbath would obviously suffer considerable material loss, while Sabbath-observing school-children would clearly have to miss one day of school every week.[64]

None of this, of course, would have even been a problem were Sabbatarians able to preserve the fixed association of the Sabbath with Saturday and of the Lord's Day with Sunday. Given the establishment of annual "blank" days, however, that would have clearly interfered with the traditional Sabbatarian obligation to observe the Sabbath precisely every seven days with no exception whatsoever. The whole essence of the Sabbath and the Lord's Day is that they are the fixed, steadfast pivots of the Jewish and ecclesiastical weeks, and the very idea of a "nomadic" Sabbath or a "floating" Lord's Day would have been sacrilegious. When Pope Gregory XIII reformed the Julian calendar in 1582, he went as far as to actually remove ten days from the calendar, yet even he did not dare to tamper with the continuous flow of the week, and arranged for Thursday, October 4 to be followed by Friday, October 15, instead of Monday![65] The continuity and absolute regularity of the seven-day week (which is a function of its having been dissociated from natural rhythms such as the lunar month and the solar year) is by far its most distinctive structural characteristic. Therefore, in principle, there is no basic difference between the interruption of the continuous flow of this cycle once every year, through the introduction of annual "blank" days, and that of the quasi weeks that we examined earlier once every month. (Interestingly enough, when the seven-day week was restored to France, the noted mathematician and astronomer Pierre Simon Laplace pointed out that this cycle can "circulate without interruption through centuries" precisely because it is dissociated from both the month and the year. By contrast, he noted how the introduction of the five "blank" *sansculottides* once every year essentially disrupted the association of any regular routine with particular days of the French Republican *décade.*[66])

The above attempts to solve the problem of the 1.2422-day "calendrical leftover" in excess of the fifty-two weeks of the calendar year are somewhat analogous to the attempts to incorporate "irrational" musical intervals into a perfectly symmetrical, "rational" system of tonal physics. In the domain of calendars, however, this "rationalist" spirit has so far faced a remarkable traditionalist opposition. The battle over the International Fixed Calendar and the World Calendar, just like those over the French Republican and Soviet weeks, was essentially a struggle between traditionalism and modernism. The fact that none of these calendars has ever been put into effect seems to demonstrate once again the power of tradition or convention in general and of religion in particular. Ignoring all the "rational" arguments brought forth and endorsed by both science and business, we have so far stubbornly stuck to the essentially "irrational," religiously based notion that the continuous flow of the seven-day week should never be interrupted. While large parts of our sociocultural environment have been "rationalized" during the past few centuries, the uninterrupted, continuous seven-day cycle remains to this day one of the most resilient "irrational" cornerstones of modern civilization.

CHAPTER FIVE

Living with the Week

A Circle in Time

The musical metaphor of the weekly *"rhythm,"* to which I have often referred throughout this book, captures the very essence of the phenomenon "week." Obviously, it presupposes a *circular conception of time.*

We often view time in the form of a line, a sort of arrowlike vector along which historically unique events are arranged in an irreversible order. This linear conception of time underlies our basic approach to history—the Jewish, Christian, and Mohammedan chronological eras, for example, are all based on historically unique events (the creation of the world, the birth of Christ, the flight of Mohammed from Mecca) that are not expected to ever repeat themselves. Yet time can also be viewed in the form of a closed circle. Essentially nonhistorical, a circular view of time revolves around the experience of *recurrence.*[1] It highlights the way in which classes of events (for example, breakfasts, geography classes, birthdays, Olympic games) repeat themselves, and involves some notion of *cycles* (the etymology of which indicates both circularity, as in Greek, and repetition, as in Hebrew).

These two modes of conceptualizing time are not necessarily

mutually exclusive,[2] and one can very well view time in both a linear and a circular fashion. When Jews, for example, upon lighting the Hanukkah candles, bless God for the miracles he performed "in those days at this time" (*bayamim ha-hem bazman ha-zeh*), they evidently seem to associate the festival with both a particular event in history ("in those days") and a nonhistorical point within the calendar year ("at this time"). Likewise, locating an instant within "1981" along the arrow of historical time does not prevent us from also designating it as "11:19 A.M., Thursday, November 26," and thus locating it within four different wheels (the daily, weekly, monthly, and annual cycles) that roll along that historical road (see Figure 11).

The circular conception of time underlies the way in which we structure our life around the week: "The concept of the week is a concept that protects you from the frightening truth that the sequence of days is not circular at all, but linear."[3] When organizing our life along the weekly cycle, we no longer seem to be thinking in terms of historical days. Rather, we think in terms of seven nonhistorical types of days that recur regularly according to a fixed periodicity.

FIGURE 11   The Linear and Circular Conceptions of Time

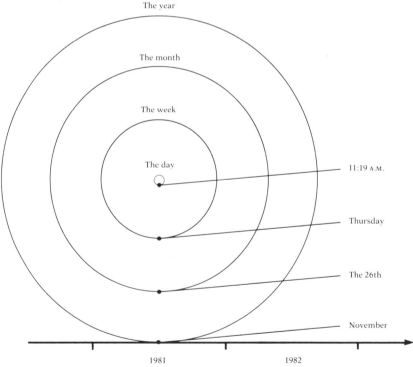

That is why we often talk about "Saturdays," yet never about "July 15, 1967s." Unlike the historical day July 15, 1967, which is not expected to ever repeat itself, Saturday is a nonhistorical type of day that is essentially experienced as one and the same entity that keeps recurring periodically.

Note also, in this regard, that Jews normally refer to Saturday as "the Sabbath." This serves to indicate that, regardless of whether it fell in April 1936 or in July 1883, the Sabbath has traditionally been experienced as essentially one and the same entity that keeps recurring periodically. Being, after all, the weekly commemoration of God's rest following the creation of the world, it is a classic example of those commemorative festivals which, according to Mircea Eliade, serve to reactualize mythical pasts in a nonhistorical "eternal present."[4] As a weekly commemoration of the Resurrection of Christ, the Lord's Day is another notable example of the reactualization of a mythical past in a sort of "eternal present" through the periodic weekly observance of a nonhistorical type of day.

Being based on the examination of annual festivals of religious calendars,[5] Eliade's theory can definitely help explain the origin of the Jewish and ecclesiastical weeks, both of which evolved within the domain of religion and essentially revolve around the periodic reactualization of religiously significant events. However, it is of very little use when we try to explain the origin of the Roman and West African market weeks or the preservation of a modified, civil version of the Judeo-Christian week in the pronouncedly antireligious French Republican and Soviet calendars. Such weeks serve to remind us that circularity is not synonymous with religion and that a circular conception of time also seems to underlie many nonreligious manifestations of the weekly cycle. In order to explain their origin, however, it is important to note that circularity entails not only nonhistoricalness but also regularity, and that a circular conception of time allows not only for the reactualization of mythical pasts in an "eternal present" but also for the establishment of regular routine.

The week is not an inevitable natural necessity and, despite its pervasive presence in so many parts of the world, it is by no means a universal phenomenon. This apparently "indispensable" cycle can actually be found only in those civilizations that either generated a complex divinatory system (for example, the Hellenistic world, Central America, Indonesia); developed a market economy (for example, China, ancient Rome, West Africa); or have come under the influence of Judeo-Christianity or Islam with their distinctive extranatural liturgical cycles. The significant feature that all these civilizations seem to have in common is that they cherish *regularity*, and it is thus only they that have produced the particular mentality which seems to characterize the sort of *"homo rhythmicus"* who has invented the cycle known as the week.

Essentially revolving around the experience of recurrence, the circular conception of time encourages the establishment of rigid *routines*, which promote structure and orderliness by making our life more regular as well as more predictable.[6] It is basically in order to protect themselves from structurelessness and irregularity that people often subject themselves to rigid self-imposed weekly routines such as going to the supermarket "every Monday" or doing the laundry "every other Friday." Likewise, establishing routine Sunday family reunions or Wednesday tutorials is probably the most effective way to guarantee that family members or professors and students will meet regularly without having to rely on ad hoc arrangements that would be necessary otherwise.

There are many routine activities (such as shopping and family visiting) that we would like to perform on a regular basis, yet which, unlike brushing our teeth or changing our underwear, need not necessarily be performed every day. At the same time, however, we also may not want to have to wait a full month in order to perform them regularly rather than on an ad hoc basis. Nature, which has given us both the day and the month, has not been of much help in our search for some convenient cycle in between them. (Unlike several other planets, Earth has no other, closer satellites with shorter orbits than the moon.) Faced with the disappointing lack of any major natural cycle that is longer than the day yet shorter than the next available natural cycle, which is almost thirty times longer, various civilizations, quite independently of one another, have tried to fill this natural gap by inventing such a cycle themselves. Hence the evolution of various forms of the week in so many parts of the world.

The Workweek

The invention of our own seven-day week essentially boiled down to the establishment of a *weekly work/rest rhythm*, based on a periodic abstention from work once every seven days. In this regard, note that in a number of languages—for example, Russian [*nedelya*], Serbo-Croatian [*nedelja*], Latvian [*nedēļa*], and Estonian [*nädal*]—the word "week" is literally synonymous with "nonwork.") As we shall now see, to this day, the seven-day "beat" is still most evidently felt within the domain of work.

Among hunter-gatherers and peasants, all temporally regular

work patterns are basically natural, stemming from rhythmic fluctuations of environmental conditions—daily variations between day and night in the quality of vision, seasonal variations in temperature or the amount of rain, and so on. Modern work patterns, by contrast, entail a considerable element of temporal regularity that is essentially artificial.[7] The artificiality of the modern work/rest "beat" results from the fact that it is dissociated from most natural rhythms, being largely derived from what Lewis Mumford characterized as a "mechanical periodicity."[8] The invention of the mechanical clock definitely played a major role in promoting artificial temporal regularity at work.[9] Yet it was the establishment of the *workweek* at least 1,500 years earlier, that allowed workers to rest on a regular basis, though according to the man-made calendar and schedule rather than nature, that had made it possible in the first place.

Particularly since the Industrial Revolution, which played a crucial role in pulling human beings away from nature, the week has been gradually replacing the year in significance, becoming second only to the day as the major cycle regulating work rhythms. It is the *weekly work schedule,* which normally involves working for five or six days in a row and then resting for a day or two, that is most responsible for providing our work with a temporally regular structure. In the United States, for example, the adjective "Monday-through-Friday" (just like "nine-to-five") is often used to describe the "normal" working conditions, as workplaces (or schools, in the case of the younger generation) usually close on weekends, and most Americans are professionally active only during the five so-called "weekdays." (The fact that both Labor Day and Memorial Day are legally fixed on Mondays and that all clock adjustments to and from daylight saving time take place at 2:00 A.M. on Sundays[10] ought to be appreciated within this context.)

Consider, in this regard, the hospital. Being morally committed to the principle of providing "continuous coverage,"[11] it operates on a continuous basis, seven days a week, which clearly ought to preclude any significant evidence of a weekly "beat." A close examination, however, reveals that, in actuality, even hospitals are only partly active during the weekend.[12] On Saturdays and Sundays, not only are most outpatient clinics closed; even inpatient units are usually covered only by a skeleton medical and nursing staff. Furthermore, during the weekend, only relatively little can be done with some patients, as various consultation services are not available and certain laboratory tests are virtually impossible to order. As a result, Friday and Saturday admissions often entail a somewhat longer hospital stay,[13] and private physicians who wish to begin treatment with no weekend interruptions often opt for scheduling elective admissions for early in the week. Consequently, on many wards, Saturday is generally regarded as a relatively "quiet" day, with most admissions being emergency cases, whereas Monday is notorious for being the busiest admission day. Since the advent of induced labor, that has also been true of many maternity wards. Newborns may not be familiar with the calendar, yet they are nevertheless born at disproportionately low rates during the weekend, and particularly on Sundays.[14]

Whether one works for six days and then rests on the seventh or for five days and then takes off the following two, workdays and working hours are normally counted today within the context of a weekly seven-day cycle. Thus, when working conditions are presented in such terms as "3/36" or "4 days, 40 hours,"[15] everyone is expected to understand that that refers to working three or four days, and thirty-six or forty hours, per week. With the main exception of pieceworkers and freelancers, workers' professional commitments are normally defined in terms of number of hours or days per week, rather than months per year or even hours per day. (To take, for example, the particular case of "flexitime" arrangements, workers do not necessarily have to put in eight hours of work on each of the five days they work, as long as they put in forty hours each week.[16]) That is also true of part-time workers, whose professional commitments are usually defined in such terms as "fifteen hours a week" or "on Tuesdays and Thursdays."

On the surface, the two days that full-time workers usually take off every week are quite indistinguishable from other 24-hour periods on which they are legitimately absent from work—holidays, vacation days, and sick days. Nevertheless, on time sheets, they are always kept separately. Thus, for example, in a hospital I once studied, nurses actually worked during Labor Day week only four days instead of the usual five, yet only two of the remaining three days were marked on their time sheets as days off, while the third one was marked as a holiday. Likewise, when they were on vacation, only five days were marked each week as vacation days, while the other two were marked as days off. Similarly, when one particular nurse was sick for several weeks, she was officially regarded as being ill only on five days each week, while the remaining two were marked as days off.[17] (The category "ill" was obviously used not to describe nurses' actual state of health, but, rather, just like the categories "holiday" and "vacation," as an official account of their absence from work. Days that had originally been scheduled as days off required no further account of the absence and therefore did not have to be marked as sick days.) In general, sick days (or holidays and vacation days, for that matter) and days off were marked separately on time sheets, because they were anchored in different "counting cycles"[18] and thus constituted entirely separate conceptual categories. Whereas the former were all granted on the basis of a certain number of days per year, the latter were all granted on a weekly basis. Being anchored in different "counting cycles," the very same phenomenon, namely legitimate absence from work, can evidently be defined as several rather distinct social realities.

Working on a "weekly" basis does not necessarily imply resting during the weekend. The ancient Jewish communities that invented the Sabbath were largely integrated through "mechanical" solidarity, which is based on likeness,[19] and the coordination within them was therefore manifested in the form of temporal symmetry. Modern society, on the other hand, is integrated much more through "organic" solidarity, which is based on complementary differentiation through a division of labor, and the coordination within it therefore often takes the form we have already examined with regard to the Soviet *nepreryvka* and the West African market week, namely, temporal complementarity. This means that not all those who work on a weekly basis necessarily take the same days off. Thus, for example, in order to service those who work Mondays through Fridays, many employees of nightclubs, museums, and department stores usually must work on Saturdays and/or Sundays, and take off one or two weekdays instead. Likewise, in the Israeli kibbutz, "although most chaverim have a rest day on Saturday, many, for a variety of reasons, have their shabbat on some other day of the week. Since certain kinds of work must be performed on Saturday. . . . those who perform these tasks have their shabbat on a weekday."[20] The notion of the Sabbath, traditionally synonymous with—and, thus, conceptually inseparable from—Saturday, has thus come to be associated with any day of the week which one takes off, as in "you have already had a Sabbath on Tuesday."[21]

While observant Jews obviously consider this traditionally inconceivable distinction between the Sabbath and Saturday sacrilegious, the kibbutzniks who take off their "Sabbath" on days other than Saturday nevertheless preserve the traditional Jewish principle of working and resting on a weekly, seven-day basis! People who rest on Saturday and those who take off Tuesday obviously structure their lives around two entirely different weekly cycles. In fact, as the Soviets have demonstrated with the continuous workweek, it is quite possible to completely abolish the common societal weekly days of rest and organize the work force around seven distinct weekly cycles that would be staggered vis-à-vis one another and revolve around seven different weekly days of rest.[22] And yet, even those seven cycles would still be only different manifestations of a single basic seven-day weekly rhythm.

Working on a "weekly" basis does not necessarily imply taking the same days off every week. Consider, for example, the pattern of working for ten days in a row and then combining one's two weekly days off with those of the following week so as to have an uninterrupted four-day "long weekend."[23] Consider also night-duty schedules whereby physicians stay at the hospital on Thursday only in week I of a three-week cycle, on Tuesday and Friday in week II, and on Monday, Wednesday, Saturday, and Sunday in week III.[24] Finally, note how some organizations rotate workers between Monday–Tuesday–Friday and Wednesday–Thursday–Saturday workweeks over a four-week period or schedule four-day and five-day workweeks on alternate weeks.[25] On the surface, all these schedules, just like those of offices that hold regular business meetings "every other Monday," seem to defy the basic weekly work rhythm, as they include no two consecutive seven-day blocks that are structurally identical. In all of them, however, an overall weekly rhythmic pattern that cannot be identified at the level of each particular seven-day week is nevertheless quite evident at the level of fourteen-day, twenty-one-day, or twenty-eight-day cycles, which are obviously precise multiples—and, thus, mere extensions—of the seven-day week. In the above hospital schedule, for example, the number of nights physicians stay on duty obviously varies across weeks, and, in any given week, some interns' night-duty load is considerably heavier than that of others. Through "averaging,"[26] however, those differences disappear at the level of the three-week cycle, within which every physician always stays on duty seven nights. The fact that, in all the above schedules, the interweekly balance is always maintained at the level of 14-day, 21-day, and 28-day—rather than, say, 11-day or 16-day—cycles is most indicative of the deliberate effort to preserve the basic seven-day weekly rhythm of work.

The introduction of periodic, weekly days of rest also affects the quality of work done on the two days that open and close the workweek. Consider, first, Monday. While some people feel particularly rejuvenated following the two-day rest, many others seem to need a few hours of "warming up" before they can break in and resettle into a productive state of mind. Given this slow start, the quality of their work on Monday often falls behind that of other weekdays, and, when something goes wrong at work, it is not unusual to hear people comment that, after all, it is a "typical Monday morning." (In France, I am told, bad cars are often referred to as "Monday products.")

The temporal proximity to the weekend also affects the quality of work done on Friday. As the last day of the workweek, it is a day on which many people succumb to the fatigue they had accumulated during the preceding four days, and allow themselves to decrease their tension and experience a weekly letdown. (By the time they get home, they often suffer from total exhaustion, regardless of the actual amount of work they had done that day. In such a physiological and mental state, it is not unusual for people to collapse, for example, in front of their television set and watch hours of programs they would never watch on any other night of the week.) This relaxation, as well as the preoccupation with how the coming weekend is going to be spent, makes Friday a most unproductive day. On Friday afternoon (definitely the ideal time for informal office parties and wine-and-cheese get-togethers), the atmosphere at many a workplace is considerably more relaxed than usual. People often leave early without expecting any repercussions, and many of those who stay usually slow down, engaging in long nonbusiness conversations with one another. Few organizations ever schedule regular business meetings for Friday afternoon.

Given that work plays such a major role in our life, we should not be surprised to find out that the institutionalization of the workweek has also affected the temporal organization of various activities that are not directly work related. Generally speaking, the weekly work/rest cycle has spurred the establishment of several complementary, mirror-image seven-day cycles that regulate activities that technically can be performed only when one is not at work. Obviously, since most of the working population is at work on weekdays, these activities usually concentrate on the weekend.

Shopping is one such activity (as are housecleaning, laundering, and lawn mowing for anyone other than full-time homemakers), and a *weekly shopping cycle*, normally peaking on Saturday,[27] seems to indicate that, in general, while the dominant motif of the weekdays is production, that of the weekend is, in a complementary fashion, consumption.[28] While it is quite possible to go to most supermarkets, for example, early in the morning or late in the evening, most of our shopping nevertheless takes place during the weekend. Traditionally, in many societies, this activity was usually associated with Sunday. (In fact, in both Hungarian [Vasárnap] and Turkish [Pazar Günü], the name of that day literally means "market day.") Today, however, throughout the urban and suburban modern world, stores normally close on Sunday, so that most shopping takes place on Saturday. It is thus quite easy to recognize that day by the large crowds that fill shopping malls, supermarkets, and department stores.

Like shopping, recreational activity too can technically take place only when one is not at work, and the weekly work/rest cycle has thus also spurred the establishment of a complementary, mirror-image *weekly recreational cycle* which peaks on the weekend.[29] The exceptionally large Sunday newspaper, the amount of time children spend watching Saturday-morning cartoons on television,[30] the fact that movie theaters usually change the films they show on Friday, and the scheduling of most spectator sports are all reflections of this cycle. The entire entertainment industry is largely organized around the weekly recreational cycle, and, throughout theaters, stadiums, museums, and discotheques, the weekend constitutes, at the level of the week, what "prime time," "rush hours," and "peak seasons" constitute at the level of the daily and annual cycles. Going out on a weekend night may be particularly costly, as tickets for shows are often more expensive, and cover charges at bars featuring entertainment considerably higher, than on ordinary weekdays. It may also preclude much spontaneity (as baby-sitters are much harder to get and reservations often need to be made a long time in advance) and involve traffic jams, long periods of waiting in lines, and crowded seating conditions. Despite all these inconveniences, however, consumers of entertainment nevertheless tend to go out on weekends much more than on any other day of the week. Monday, by contrast, is probably the "slowest" day of the week throughout the entertainment industry. On that day, when many consumers of entertainment still "recuperate" from the weekend, many theaters and museums are closed, few jazz clubs offer any program, movie theaters may try to attract audiences by charging considerably lower admissions, and Johnny Carson fans must do with repeat shows.

The weekly recreational and shopping cycles also affect the temporal organization of television programming. It is precisely the fact that they are the most popular nights for going out that makes Friday and Saturday the worst possible viewing nights from the standpoint of advertising sponsors. Instead of "wasting" their high-rated shows, major releases, and most successful movies on those nights, television networks would rather air them, for example, on Thursday, a big viewing night which is sponsors' last—and, thus, possibly also the most effective—opportunity to affect those whom they regard as "big spenders" before they get their weekly or biweekly paychecks and set out on their weekend spending sprees.

Major releases and successful movies are also aired on Sunday night, which advertising sponsors regard as the ideal time for attracting the widest possible range of age-groups, since it is a time when families often watch television together. This seems to imply some *weekly cycle of socializing,* which has indeed also evolved as a mirror image of the work/rest cycle. Emile Durkheim, who first noted how social life oscillates in accordance with certain rhythmic patterns, attributed this intermittent character of society to the calendar, the foremost function of which he considered to be the regularization of the periodic reassembling of social groups.[31] The week plays a major role in the regulation of such "social traffic," as it constitutes the foremost cycle of social concentration and dispersion.

It is by no means a mere coincidence that, when Japan launched its surprise attack on Pearl Harbor in 1941, it chose to do that on a Sunday morning, and that was probably also true of the Polish government's decision, in 1981, to crush the Solidarity movement on a Saturday night. Consider also, in this regard, the bargain rates— normally associated with using services at times other than peak demand periods[32]—that telephone companies and airlines often offer their weekend customers, as well as the scheduling of all clock adjustments to and from daylight saving time for 2:00 A.M. on Sunday. All of the above reflect the basic pattern whereby most "business" interaction is normally suspended during the weekend.

At the same time, however, in a somewhat complementary fashion, precisely during those parts of the week when work-related interaction reaches its lowest ebb there is the sharpest increase in active social interaction within families and among friends. Like shopping and recreation, socializing is also an activity for which we have more time when we do not have to be at work. The weekly cycles of work and socializing thus essentially complement one another, and the social "density"[33] of family life and friendships can actually be plotted along fairly regular weekly rhythmic patterns, with the highest volume of interaction obviously taking place during the weekend.[34] That all this is much less true of the very young and very old,[35] who are not an integral part of the work force, further attests to the intimate interrelatedness of the weekly cycles of work and socializing.

On weekends, married adults usually minimize their individual involvements and engage in social activities almost exclusively as couples. These are also the days on which city dwellers make family trips to the country[36] or the beach. For children, weekends are the best days for spending long periods of time with their parents—in bed before breakfast, in the park, attending ball games, going to the matinees, and so on. In fact, they are the only days on which some families ever get to spend any time together. Consider, for example, dual-career couples who work in different cities, or broken families, where the "absent" parent is periodically available only on weekends, as in the classic case of the so-called "Sunday father." The weekend is also the time for family visits, as well as the time when most letters are written[37] and telephone calls made to relatives who live in other cities or countries.

Fridays and Saturdays are traditionally also the big nights for dating. Likewise, parties are given and friends are visited almost exclusively during the weekend. In part, this stems from the fact that, on working days, people are usually more tired and also have much less time for such preparations as cooking and cleaning, which entertaining guests normally entails. However, the fact that more "social evenings" are scheduled for Fridays than for Sundays seems to indicate that having them on nights immediately preceding nonworking days is at least as crucial as having them on nonworking days. Not having to work the following morning allows one to be more relaxed in general, refrain from glancing too often at one's watch, drink more, and stay up later. Being the only night that both follows and immediately precedes nonworking days, Saturday night is obviously the most popular time for sociable activities.

The more relaxed atmosphere of Friday and Saturday nights parallels that of Saturday and Sunday mornings. Not only do we go to sleep later than usual on Saturday nights, we also wake up later than usual on mornings when we do not have to go to work, and that is particularly true of Sundays.[38] These weekly rhythmic sleep patterns clearly indicate that even purely biological functions are often regulated not only by internal organic rhythms, as biologists would have us believe, but also by entirely conventional social rhythms such as the week. This is also true of our weekly rhythmic eating patterns, all of which derive from the weekly cycles of work, shopping, recreation, and socializing.

Note, first, the weekly variations in the amounts of food we consume. As the most popular time for dinner parties, family visits, and evenings at restaurants, all occasions when we usually eat in a considerably less controlled manner than we normally do, the weekend is clearly the most difficult time for adhering to a strict diet. (For similar reasons, it is also the part of the week when people consume the largest amounts of alcohol, which may help explain why Saturday, followed by Sunday and Friday, has the highest rates of homicide as well as death caused by motor vehicle accidents.[39]) Second, there is a weekly variation in the amount of choice we have regarding what we eat. As most grocery shopping takes place during the weekend, Saturday is usually the day when our refrigerator and cupboard offer us the greatest amount of choice, whereas on Thursday, by contrast, our food supply has usually dwindled to the point where our degree of freedom in choosing often approaches zero. Finally, consider the weekly variations in the quality of our meals. Weekday meals (particularly in households where both spouses work outside home) usually consist of food that does not take long to prepare. This applies even to breakfasts. Contrast, for example, the ordinary weekday breakfast with the weekend breakfast, a classic example of which is the Sunday brunch. Given the pressure to leave home on time for work, weekday mornings usually entail only "abbreviated" versions of such activities as going to the bathroom and eating breakfast, and the latter is often even skipped altogether. Weekend mornings, on the other hand, usually open on a much more relaxed note and proceed at a relatively slower pace. (Even those who compulsively turn on the radio "first thing in the morning" may fail to do so on weekends.) Weekend breakfasts therefore usually begin somewhat later than usual,[40] take longer to prepare, and often last much longer. The very same people who are often too much in a rush to finish drinking even one cup of coffee on ordinary weekday mornings usually drink several cups at a rather leisurely pace on Sunday brunches.

Thinking in Weeks

Thus, the weekly structuring of work also affects the temporal organization of activities that are only indirectly related to work. However, the weekly work/rest cycle does not help explain why *Newsweek* is published precisely every seven days, why particular yoga classes are offered only on Wednesdays, or why Tuesday is the day on which Manhattan's Bleecker Street Cinema once offered only Japanese films and the English used to have their public executions.[41] Nor can it account for the fact that English country fairs and American civil holidays are regularly fixed on such dates as "the first Wednesday following February 2" (the Hereford fair), "the fourth Thursday in November" (Thanksgiving Day), "the Thursday immediately preceding the second Friday in October" (the Leicester fair), or "the Tuesday next after the first Monday in November" (Election Day).[42] Consider also regular Wednesday business meetings, Monday chemistry labs, Friday therapy sessions, Tuesday television shows, and regular Thursday lunches with particular friends. It is precisely such weekly patterns and routines, none of which can be explained by the weekly work/rest cycle, that give our life its distinctive seven-day "beat."

The phenomenon, incidentally, is not peculiar to modern civilization. In ancient Judea, for example, courts used to be held only on Mondays and Thursdays, Levites at the Temple would recite different hymns on different days of the week, and maidens and widows could get married only on Wednesdays and Thursdays, respectively.[43] All this shows that even two thousand years ago the week was already much more than just a seven-day work/rest cycle.

The only way to account for all the above weekly patterns and routines is to see them as manifestations of the fact that we think in terms of weeks. With the weekly work/rest cycle affecting so much of our everyday life either directly or indirectly, we have become habituated to thinking about the passage of time—and, thus, also to measuring it—in terms of seven-day units. (For example, in Walt Disney's adapted version of Sir James Matthew Barrie's famous play, when Peter Pan decides to banish Tinker Bell from Never Never Land forever and Wendy protests, the immediate compromising solution he comes up with is "for a week, then.") Consequently, we have also come to structure even those parts of our life that bear no relation whatsoever to work along weekly patterns.

Evidence of the establishment of the week as a standard unit for measuring the passage of time can be found in the Old Testament more than two thousand years ago:

> Seven weeks shalt thou number unto thee; from the time the sickle is first put to the standing corn shalt thou begin to number seven weeks.[44]

> In those days I Daniel was mourning three whole weeks. I ate no pleasant bread, neither came flesh nor wine in my mouth, neither did I anoint myself at all, till three whole weeks were fulfilled.[45]

This incorporation of the week into a mathematical conception of time, whereby intervals that are perceived as quantities of abstract duration are used as standard units for measuring the passage of time,[46] was a major intellectual breakthrough that essentially involved transcending the original Jewish conception of the week as a seven-day cycle that begins on the day immediately following the Sabbath. The week was to be conceptualized as a series of any seven consecutive days, that is, as a seven day interval that can begin at any point in time.

The measurement of the passage of time in terms of weeks is manifested, for example, in the way both parents and pediatricians normally reckon babies' age. Even more spectacular, however, is the way in which the period of gestation is reckoned, as, for example, in legal regulations that restrict abortion beyond the completion of a certain number of weeks of pregnancy. These regulations seem to indicate that even the age at which a fetus is first regarded legally as a living human being is actually set not according to any natural events such as conception or birth, but, rather, according to a social calendar that is essentially based on extranatural, conventional units of time such as the week.

Thinking about the passage of time in terms of weeks normally goes hand in hand with organizing one's life along weekly patterns. Thus, for example, given that much of their hospital stay is often structured along the weekly cycle,[47] it is quite understandable that patients would reckon the passage of time in hospitals in terms of weeks, as any reader of Thomas Mann's classic *The Magic Mountain*[48] must know. Likewise, given that in hospitals work is normally orga-

nized along weekly patterns and routines whereas the month is somewhat irrelevant, it should come as little surprise that doctors would
regard 31-day and 45-day rotations as being "four weeks" and "six
weeks" long, to the point of even correcting someone referring to
the latter as being a month and a half long.[49]

Through a simple mental act of multiplication, we have managed
to introduce into our life *multiple-weekly cycles* that are essentially
precise multiples of the seven-day week. Consider, for example, six-
week tennis or arts camps, four-week exercise and dieting programs,[50]
fourteen-week college semesters, and three-week bank-tellers training
programs. Also similar in principle are athletes' training schedules,
whereby seasons are essentially divided into multiple-weekly blocks
which constitute the elementary units along which training, peaking,
and resting are temporally organized.[51] Consider, for example, a run-
ner's thirty-nine-week annual training schedule, which consists of
various eight-week, six-week, four-week, three-week, and two-week
blocks, each of which is characterized by a distinctive training
program.[52] Consider also, in this regard, the way some high-jumpers
organize their weight-lifting training in sixteen-week blocks, each
one consisting of four distinct four-week cycles.[53]

Multiple-weekly cycles are not peculiar to modern civilization
alone. Note, for example, the ubiquitous presence of 26-day, 39-day,
52-day, 65-day, 78-day, 91-day, 104-day, and 117-day periods in Maya
manuscripts,[54] all of them precise multiples of the ancient Central
American thirteen-day week. I have also mentioned earlier the West
African eight-day and sixteen-day market cycles which are believed
to have derived from the basic four-day market week. Finally, note
the major role played by the *trinundinum* (or *trinum nundinum*),
an interval involving three consecutive market days, in various Roman legislative and electoral procedures. It was the minimum waiting
period that had to elapse, for example, between the introduction of
candidates for public offices and the actual elections, or between
the promulgation of proposed laws and their presentation in the form
of draft bills before the assembly.[55]

Ever since the introduction of the oral contraceptive commonly
known as the "pill," conventional multiple-weekly cycles have been
asserting their pervasive presence even within the realm of the human
body. Most pharmaceutical companies package the pill in twenty-
one-tablet or twenty-eight-tablet sets,[56] which entails a pattern of taking the estrogen-containing contraceptive for twenty-one consecutive
days and then stopping (in the case of the twenty-one-tablet sets)
or taking mere placebos or tablets containing iron and vitamin sup-
plements (in the case of the twenty-eight-tablet sets) for the following
seven days. Since not all women have a twenty-eight-day menstrual

cycle, the rationale behind this 21–7 pattern cannot be entirely physiological. The only advantage it has over a 21–6 or a 20–7 pattern, for example, is *cognitive convenience*, since it helps users form a routine habit of always starting a new package of pills on the same day of the week: "Most women now take their pills on 21 and 7 day schedules. There's very little difference between this and a 20 and 7 or 21 and 6 day schedule but a full 28 days allows a woman to begin taking her pills on the same day each month. Having to remember "Tuesday" is a lot easier than counting days all the time."[57] That also helps to explain the packaging of progestogen-containing "minipills" in thirty-five-tablet or forty-two-tablet sets,[58] as well as the fact that regular twenty-one-tablet and twenty-eight-tablet sets are often arranged in groups of seven tablets. After all, twenty-one-day, twenty-eight-day, thirty-five-day, and forty-two-day cycles are all precise multiples of the basic seven-day weekly cycle. What essentially takes place here is that the perfectly natural menstrual cycle (which is not always twenty-eight days long for all women) is being replaced by artificial cycles that are all mathematical extensions of an entirely conventional social cycle. Rather than adapt the temporal pattern of taking the pill to their own internal rhythms, users normally adapt their own physiological makeup to the entirely social habit of thinking in terms of seven-day intervals and their precise multiples.

It is also for the sake of cognitive convenience that people sometimes arrange to have their house cleaned "every third Thursday" and that employers often pay their workers "every other Friday." Consider also a routine such as having a regular open house on the first Monday of every month. From the standpoint of cognitive convenience, whether any two consecutive occurrences of this event are spaced twenty-eight or thirty-five days apart from one another is relatively inconsequential, as long as they are spaced along intervals that are *precise multiples of the week*. That allows the event to always fall on—and to be associated in people's mind with—the same day of the week. Along similar lines, when doctors want to make sure that patients remember to come to the clinic on a regular basis, yet slightly less frequently than "every Tuesday," they usually do not even consider establishing a pattern such as "every nine days" and normally schedule their appointments for "every other Tuesday." Similarly, when people want to call their family on a regular basis, yet not as often as every week, they usually ignore other possible regular patterns such as "every eight days" and, essentially halving the "density" of the relationship, establish a routine of calling them "every other Sunday." In other words, the choice is not so much between calling every seven or eight days as between calling every seven or fourteen days, that is, on a weekly or a biweekly basis.

Our tendency to structure our life along patterns that are based on cycles that are precise—rather than mere approximations of— multiples of the week may also account for the fact that beach houses are usually rented on a weekly, rather than a daily, basis and that workers are sometimes required to take their vacations in weekly or precisely multiple-weekly segments.[59] As a unit of time, the week is often handled as if it were indivisible, which seems to exemplify Henri Hubert's original claim, nicely paraphrased below by Pitirim Sorokin, regarding the perceived indivisibility of social calendrical units:

> Many periods of sociocultural time, sometimes of a long mathematical duration and easily divisible mathematically, are socially indivisible. They are living units. When a quantitativist breaks such units into a certain number of seconds, hours, or other equal units, he breaks the living unity of this time portion, distorts the reality, and loses the nature, "wholeness," "meaning," and *"Gestalt"* of such periods.[60]

Given the total absence of any standard unit of time between the week and the month, the most convenient way to structure the regular recurrence of activities less frequently than every week yet not as infrequently as every month is to organize them along multiple-weekly patterns such as "every other Saturday" or "every third Wednesday." Just imagine, for example, trying to sustain on a regular basis nonweekly routines such as doing our grocery shopping or laundry every twelve days or getting our periodic haircut every twenty-two days. Such routines are doomed to fail because, given the unavailability of any standard unit of time that corresponds to them, adhering to them on a regular basis would obviously present a formidable cognitive inconvenience. Essentially involving rhythms that would go against the dominant seven-day "beat," they would only confuse us. That also explains why there are no magazines that are published on a regular basis every seventeen days, business meetings that are regularly scheduled for every eleven days, and radio programs that are regularly aired every eight days.

For the very same reason, given the total absence of any standard unit of time between the day and the week, it would also be most inconvenient, from a cognitive standpoint, to establish nonweekly routines that involve cycles that are longer than the day yet shorter than the week. Taking a piano lesson every six days or calling someone regularly every four days is at least as uncommon as doing the laundry every twelve days. People often try to wash their hair, water

their plants, change their undershirt, or alternate sets of daily exer-
cises "every other day," yet they usually find it most difficult to adhere
to such routines on a regular basis for a sustained, uninterrupted
period of time.

Adhering to "every other day," "every third day," "every fourth
day," "every fifth day," or "every sixth day" routines on a regular
basis is most inconvenient, from a cognitive standpoint, because it
necessarily involves performing the same activity on different days
on alternate weeks. (A six-day week, by contrast, would allow doing
things regularly every other day or every third day yet also on the
same days every week.) Evidently, we cherish the ability to maintain
the fixed association between particular activities and particular days
of the week to the point where we are even willing to pay the price
of having to give up the perfect mathematical regularity that does
not seem to go along with it. In other words, we would rather do
things on the same days every week than do them regularly every
two, three, four, five, or six days. Thus, ironically, in order to do
things on a regular basis, we often choose to abandon isochronal
(that is, durationally uniform) intervals altogether. Instead, we establish regular routines such as taking a class "every Tuesday and Friday"
or jogging "Mondays, Wednesdays, and Saturdays." A Tuesday–Friday
routine obviously involves constant alternations between three-
day and four-day intervals between recurrent activities. On the other
hand, it definitely allows for a regular weekly schedule.

It should become clear by now that the week functions not only
as a standard unit for measuring the passage of time, but also as a
framework within which activities can be organized in a systematic
fashion. The need for such an organizing framework is evidently
so great that when the seven-day week does not offer the best opportunity for "rational" organization, it is likely to be replaced by a different, yet nevertheless "weekly," cycle that would serve as its functional
analogue. A perfect case in point is the "day cycle"[61] established by
some American schools as a substitute for the conventional school
week, mainly in order to alleviate problems resulting from the fact
that holidays and days with heavy snow are not distributed evenly
across the days of the week, thus hurting some classes more than
others. This cycle consists only of days when the school is actually
open,[62] and classes are scheduled for days which, unlike those of
the conventional five-day school week, are not synchronized on a
regular basis with particular days of the week. (As we can see in
Figure 12, unlike the conventional school week, which always begins
on Monday, a six-day "day cycle," for example, always begins on
day A, regardless of whether it is a Monday or a Friday.) The "day
cycle" may not necessarily be seven days long and its days are deliber-

FIGURE 12 The "Day Cycle"

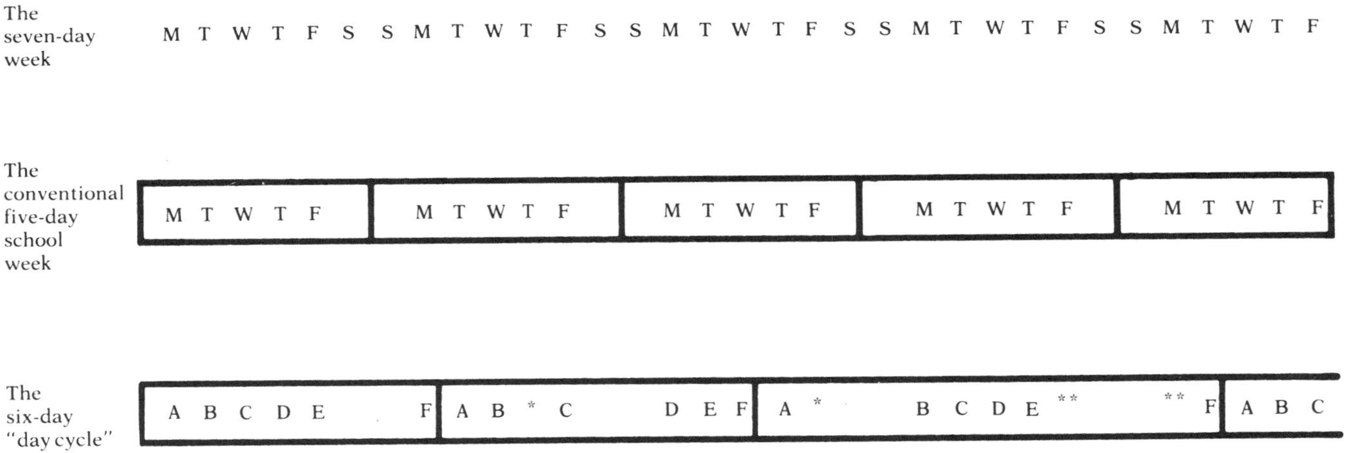
The seven-day week
M T W T F S S M T W T F S S M T W T F S S M T W T F S S M T W T F
The conventional five-day school week
M T W T F
M T W T F
M T W T F
M T W T F
M T W T F
The six-day "day cycle"
A B C D E F
A B * C D E F
A * B C D E ** ** F
A B C

* School is closed due to heavy snow.
** School is closed due to holiday.

ately not synchronized with those of the seven-day week, yet it is nevertheless a weekly cycle! Like the seven-day week, it was invented in order to fill the gap left by nature between the day and the month and allow us to organize the regular recurrence of events such as history classes and chemistry labs in the most systematic, "rational" fashion.

As an organizing framework, the week first constitutes a *"counting cycle."*[63] Thus, for example, when the Federal Communications Commission evaluates radio stations' programming performance, it counts the time devoted to news and to commercial matter, the number of public service and spot announcements, and the total number of hours of actual broadcast against a "composite week" that consists of seven broadcasting days, each one representing one particular day of the week.[64] It is also against the week that television networks count the number of times they broadcast the same program (consider, for example, "once-a-week" series) as well as the number of hours they broadcast a particular type of program.[65] Likewise, in schools, all scheduling is done within the context of the school week, and even radical reformers who defy conventional units of time such as the hour and organize their schedules along flexible ten-minute "modular time units" nevertheless abide by essentially weekly patterns such as "eleven modules per course a week."[66] Runners, too, organize their training in weekly terms, so that the variation between preseason, early-spring, and late-spring training, for example, is largely manifested in the difference between doing six, four, or three workouts per week.[67] They also count the number of miles they average per week, defining progress, for example, in such terms as a 15 percent increase in their weekly training mileage.[68]

Using the week as a counting cycle, we can totally ignore the day as a relevant reference framework, as, for example, in the case of dieting: "The point to remember is that it's the over-all weekly calorie count that matters. You can put on an extra pound and a half over a weekend and take it off by the next Friday night."[69] Thus, we can maintain a balanced diet without having to worry about maintaining an even daily calorie input, by adhering to a weekly rhythmic pattern whereby we both watch our weight from Monday through Friday and abandon ourselves to gluttony during the weekend (though such a diet perhaps is not recommendable from a medical standpoint). Such a weekly pattern also resembles the one whereby we both spend our money rather cautiously during the week and indulge in extravagant buying sprees during the weekend. At the same time, along similar lines, we can also maintain a "weekly" balance by ignoring the week itself, using instead a multiple-weekly cycle, as a counting framework. We have already seen this upon examining the practice of "averaging" weekends, days off, and nights on duty in the case of the workweek. Consider also, in this regard, the way "coparents" often organize a balanced "coverage" of their children through multiple-weekly schedules such as having them on alternate weeks, on alternate weekends, three out of every four weekends, and so on.[70]

As a counting cycle, the week obviously functions as a budgeting device. First, it constitutes a framework within which at least some minimum incidence of any particular type of event may be guaranteed. Scheduling events on a regular basis helps assure that they occur in the first place,[71] so that by establishing a routine whereby I spend an afternoon with my daughter on a regular basis every Thursday I may guarantee that I spend some exclusive time with her at least once a week. Incidentally, in order for an event to occur on a regular "weekly" basis it does not necessarily have to recur precisely every seven days (such as "every Thursday"), as long as it occurs regularly within the boundaries of every seven-day period. Thus, for example, in order to go to the swimming pool on a regular "weekly" basis, I do not necessarily have to go there "Mondays, Wednesdays, and Fridays," as long as I go "three times a week." Similarly, if I commit myself to cleaning my house on a regular "weekly" basis, I must make sure that, if on any particular week I cannot do that on Saturday, as usual, I should make a special effort to do it on some other day within the same week. Finally, even when no actual weekly routine is established, the fact that the week is used as an accounting device may in itself encourage some minimum incidence of particular activities. Thus, for example, the very fact that couples sometimes use the week as a framework for counting how often they have sexual intercourse[72]—an obvious case where it serves to fill the gap left by nature between the day and the month, since reckoning by either of those cycles would most probably prove too impractical—may indeed help them maintain some minimum frequency of that activity.

Its regulative function as a budgeting device is also manifested in the diametrically opposite case, when the week constitutes a limiting framework within which some maximum incidence of particular events is maintained. This is exemplified by the case of couples who, quite unlike children who promise to call their parents at least once a week, decide to get together no more than once a week (or "only on weekends"). Consider also, in this regard, nutrition-conscious people who establish routines such as not eating meat more than twice a week, drinking coffee only on Sundays, eating no more than four eggs a week, and allowing their children to have sweets only on weekends. Note also how, during periods of shortage, consumers are often restricted to purchasing rationed food items or gasoline only on certain days of the week. Finally, there is the use of the week as a framework for budgeting finances, by children who get from their parents a weekly allowance as well as by adults who, even when they are paid by the month, nevertheless plan to spend so many dollars per week on food and establish routines such as not going to the movies more than once a week.

The regulative function of the week as a budgeting device is also manifested in the way responsibility is allocated among individuals. This essentially involves a division of labor which, peculiarly enough, does not entail permanent functional differentiation, as all participants assume the very same responsibilities (though, obviously, at different times). Such a division of labor usually takes one of two forms, namely, a *weekly rotation system*, whereby each participant assumes responsibility for performing a particular task on a separate week for the entire week; or a *weekly schedule*, whereby each participant is responsible for performing that task only on certain days of the week, and on the same days every week. The former arrangement is exemplified by the manner in which Jewish priests would rotate service at the Temple in Jerusalem,[73] Benedictine monks would organize kitchen duties (as well as reading to the community during meals),[74] and teachers still allocate class duties. The latter arrangement is manifested in the way many married couples organize daily domestic responsibilities such as making the bed, taking the children to and from school, preparing dinner, walking the dog, and so on.[75]

In the same way that a number of individuals may alternate periodically in assuming the responsibility for performing a single task, a single individual can alternate periodically among a number of related tasks, which together constitute a single systemic whole, that he or she is responsible for performing. The following nursery rhyme, for example, suggests a possible weekly cycle of Pilgrim domestic chores:

> *Wash on Monday,*
> *Iron on Tuesday,*
> *Bake on Wednesday,*
> *Brew on Thursday,*
> *Churn on Friday,*
> *Mend on Saturday,*
> *Go to meeting on Sunday.*[76]

The same pattern can also be applied at the level of any given single type of activity. Thus, for example, in order to organize in a most systematic fashion the variability among food items consumed, a family might eat cheese, macaroni, and baked chicken on Sundays, greens, cornbread, and baked beans on Wednesdays, and so on.[77] Along similar lines, a housewife might decide to organize the cleaning of her house by cleaning a different room each day of the week on a regular basis. (In doing that, she would resemble Benedictine monks who managed to complete chanting the full cycle of 150 psalms every week by associating each day of the week with particular psalms on a regular basis.[78]) Consider also, in this regard, the pattern according to which the streets of Manhattan's Upper West Side are regularly cleaned—one side on Mondays, Wednesdays, and Fridays, and the other on Tuesdays, Thursdays, and Saturdays.

In a similar manner, athletes normally use the week as the foremost organizing framework within which they allocate their involvement in various types, as well as amounts, of training in the most systematic fashion. Following the ancient Greek model of organizing training along four-day cycles that allowed for periodic alternation among preparation, concentration, relaxation, and moderation,[79] they associate each day of the week with a distinctive training routine. Thus, they manage to systematize the periodic alternation among stress, pace, endurance, and speed workouts by associating each of those with particular days of the week on a regular basis. For example, middle-distance and long-distance runners' training schedules are, with no exception, weekly schedules.[80] Each week constitutes the fundamental framework within which they systematically organize the allocation of their involvement in continuous slow running, continuous fast running, fartlek running, repetition running, slow intervals, fast intervals, pace intervals, and sprints. They do that by associating each of those types of workouts with particular days of the week on a regular basis.

Interestingly enough, precisely along the same lines, schools normally design their curricula, using the week as the foremost organizing framework for establishing certain desired proportions among the various courses that are offered. A particular school's relative emphasis on art and mathematics, for example, is most conspicuously reflected in the amounts of time per week devoted to either subject.

Thus, the week is often used not only for establishing periodic variability among different types of activity, but also for systematizing certain desired proportions among them. To put this more generally, a regular weekly schedule allows us to allocate our involvement in the various domains of our life in a way that makes a certain desired balance among them possible.

Given that modern life usually entails a particularly fragmented mode of existence, the week seems to constitute a perfect framework for organizing in the most systematic manner its compartmentalization into several isolated domains. This essentially involves splitting this cycle into a number of distinct clusters, each one consisting of several days that are usually (but not necessarily) arranged in a single continuous block and distinctively associated with one of those domains. Part-time workers who arrange to work on Monday, Tuesday, and Wednesday, so as to also have, aside from the time they share with their families on Saturday and Sunday, a sort of "personal weekend" for themselves only on Thursday and Friday, are a perfect case in point. So are commuters who live five days in one city and two days in another, and academics who teach on Mondays and Tuesdays, write on Wednesdays, Thursdays, and Fridays, and do research on Saturdays and Sundays (or write fiction from Monday through Thursday and nonfiction on Fridays[81]).

The *weekly compartmentalization of human life* highlights the fact that the week provides our existence with some structure. Aside from imposing an unmistakable rhythmic "beat" on a vast array of our activities, it also helps to structure the actual differentiation among them. The week disrupts the otherwise continuous flow of our everyday life on a regular basis and, in doing that, adds more dimensions to our existence.

CHAPTER SIX

Experiencing the Week

W̶E VERY OFTEN THINK of time in mathematical terms, that is, as homogenous quantities of duration. From this mathematical standpoint, one hour is essentially interchangeable with any other hour, as both consist of precisely sixty minutes. However, as Henri Bergson demonstrated in his seminal work on the psychology of time, mathematically equivalent durations can nevertheless be experienced as having quite different qualitative "intensities" or feeling tones.[1] His work made us aware of our *qualitative experience of time* as a heterogenous entity.

Bergson's theory, however, was essentially psychological, and obviously lacked a sociological dimension. The necessity of the latter was made clear a few years later, when sociologist Henri Hubert pointed out that the source of the differing feeling tones of mathematically equivalent durations is very often society.[2] Thus, for example, it is only the Church that makes Lent experientially different from any other forty-weekday period within the calendar year, and it is the university that is responsible for making the examinations week "feel" like no other week in the semester. Similarly, it is American society that, by commemorating the last three decades of the eighteenth century far more intensively than any other thirty-year period in its history, manages to attach different meanings to historical periods of equal duration.[3]

The Weekend

One of the classic manifestations of humans' qualitative experience of time is the cultural differentiation among the seven days of the week. Despite their physical-mathematical equivalence, these twenty-four-hour intervals are socially regarded as seven distinct types of days with entirely different feeling tones.

"Periods of time acquire specific qualities by virtue of association with the activities peculiar to them."[4] First-graders, for example, already associate Tuesday with having to prepare their weekly mathematics home assignment, whereas women whose husbands take evening classes on that day may come to associate it with a less elaborate dinner and longer telephone conversations. Likewise, for parents whose older child has after-school activities on Wednesday, the latter may come to "mean" an exclusive afternoon with their younger child. Yet the days of the week acquire their distinctive meanings not only from the activities occurring within them, but also as parts of a larger pattern,[5] and particularly from their temporal location vis-à-vis other days. Thus, for example, for those who take regular music lessons every Thursday, Wednesday may "mean" the last day on which they can still practice their études. Similarly, for teachers who have a heavy schedule on Mondays, Sunday may "mean" a bad night for going out.

Various days often function as the principal temporal landmark or milestone within the weekly cycle, with respect to which all the other days of the cycle acquire their distinctive meanings. Consider, for example, the special significance of Thursday for lovers who meet regularly as well as exclusively on it,[6] or for,

> certain unemployed urban Negro men who bide their time until Thursday night, typically the suburban Negro maid's night off, whereupon they drive to the railroad station, pick up one of the girls, and enjoy a night of revelry and profit. The period between these Thursday nights is defined as "dead" time—a period of social inactivity.[7]

Along similar lines, those who get paid on Friday often regard that day as the pivot on which the entire week turns.[8] By the same token, for those working in a newspaper that is regularly published every Wednesday, Tuesday and Thursday may come to "mean" preparation for Press Day and cleaning after it, respectively.[9] However, all the above idiosyncrasies notwithstanding, it is basically the *weekend* that, for most of us, functions as the principal temporal milestone with respect to which all the other days of the week acquire their distinctive meanings.

Consider, for example, the way we normally experience Monday: "If a person feels blah, acts sluggishly, and cannot easily make decisions, there may not be anything deeply wrong with him or her; it may simply be Monday."[10] Monday's notorious association with gloominess and low morale is so strong in our culture that, upon warning her readers not to expect much passion and melodrama in her novel *Shirley*, Charlotte Brontë simply indicates that "something unromantic as Monday morning" lies before them.[11] The general feelings associated with this day range from anger[12] to melancholy. Cartoonist Jim Davis's cat Garfield defines it as "a day designed to add depression to an otherwise happy week,"[13] and the lover in the song "Monday, Monday" complains that, while "every other day of the week is fine," Monday mornings usually find him crying.[14]

While cartoons and popular songs often provide us with some penetrating insights into the human soul, the "hard" evidence is even more compelling. Consider, for example, the fact that almost half of all sickness spells begin on Monday and that it is the busiest day in hospital emergency rooms.[15] It is quite hard not to take a cynical stance and see this association of Monday with the highest rates of absenteeism from work[16] as a modern version of the so-called "cult of Saint Monday"[17] that has haunted employers at least since the Industrial Revolution. As cultural historian Lawrence Wright has so wittily observed, "viruses seem able to distinguish between shift workers and day workers, and can even understand the calendar. They tend to make day workers go sick on Mondays, but very few shift workers go sick on a Monday which falls in their rest period."[18] At the same time, however, it is impossible to be cynical about clinical evidence such as the excess proportion of cardiac deaths—normally indicative of some serious psychological stress—among men on Monday.[19] Even more compelling is the fact that, of the seven days of the week, Monday is clearly the one with the highest suicide rates.[20]

Only a nonmedical, sociological explanation can account for these findings. That so many people attempt suicide, suffer sudden cardiac death, call in sick, or else are depressed, irritable, and full of complaints on Mondays can be explained only by the latter's temporal location vis-à-vis the weekend. The sharp contrast between Monday and the day immediately preceding it is the most distinctive experiential characteristic of that day. Thus, for example, in the song "Sunday," a lover feels blue every Monday, when thinking over Sunday.[21] Likewise, in another song, it is essentially a "Sunday kind of love" that a lover feels she needs in order to keep her warm "when Mondays are cold."[22] Mondays are experienced as "cold" because they are associated with the transition from the attractive world of rest, playfulness, lack of responsibility, and intensive contact with loved ones to the serious, mundane, and demanding world of work. These experiential characteristics of Monday morning often rub off on parts of Sunday as well. The mere anticipation of the approaching day they dread and regard as their least favorite day of the week leads many people to experience sometimes a sort of "Sunday evening (or even afternoon) blues."

That Monday's distinctive qualities are essentially a function of its temporal location immediately following the weekend becomes even more obvious once we contrast them with those of the day immediately preceding the weekend, Friday. Not surprisingly, these two days, which share no other characteristic in an exclusive manner, are sometimes coupled, as in the following aphoristic allusion to the fundamental contrast between sybaritic hedonists and compulsive workaholics: "There are two kinds of persons—those who look forward to Friday and those who look forward to Monday." The contrasting cultural meanings of these two days are most clearly a function of their contrastive relationship to the weekend. Whereas Monday, the postweekend day, is normally associated with a dreadful sense of "there we go again, back to the same routine," Friday, the preweekend day, is usually associated with an elated feeling of "it's finally over." The complementary cultural counterpart of the notion of "blue Monday"[23] is, thus, the popular expression "Thank God it's Friday" (TGIF).[24] Not surprisingly, in contrast to the "cult of Saint Monday," Friday is the least popular day for sickness spells to begin,[25] and is normally characterized by a considerably cheerful mood.[26] (One Friday afternoon, when I asked one of our secretaries why she was looking so cheerful, she replied "it's Friday" and seemed genuinely astonished at my having asked such a silly question.)

While clearly not a part of the actual weekend (after all, it is a day on which one still goes to work or school), Friday is nevertheless considered by many their favorite day of the week, because it involves an experience many of us evidently cherish, namely, anticipation. For those of us who already as very young children often found pleasure in counting down the days remaining before the weekend and who possibly even prefer the experience of anticipating the weekend to the experience of the weekend itself, weekends may feel as anticlimactic as are homestretches to runners who have already established their victory long before the race is officially over. After all, on Friday we can experience a feeling that we can never experience on the weekend itself, namely, that there are still a couple of weekend days before us! In fact, some people therefore find Thursday, which involves the anticipation of both the weekend and Friday, even more gratifying than the latter. (The anticipated weekend, however, must be experienced as being within reach, which explains why few people, if any, consider Tuesday or even Wednesday their favorite day of the week.)

Let us not forget, however, that it is basically the attractiveness of the weekend that accounts for that of Friday. It is Saturday that D. Gates has in mind when claiming, in his song "Saturday Child," that there are seven days in a week made to choose from, but only one is right for him.[27] It is Sunday that is portrayed in the song "Sunday in the Park" as the best day, the day we keep on looking forward to all week long.[28] It is Saturday and Sunday that can also boast the lowest suicide rates.[29]

Much of the attractiveness of the weekend can be attributed to the suspension of work-related—or, for the young, school-related—obligations. Admittedly, there are some people who do look forward to Monday rather than Friday and who may even feel a particular sense of accomplishment when they manage to work during the weekend. Workaholics' compulsive need for structure is sometimes so great that they may even develop the syndrome identified by Sándor Ferenczi as "Sunday neurosis."[30] To quote Sebastian De Grazia,

> the Sunday neurosis, as it sometimes is called, seems fairly widespread in a milder form as a malady exhibiting a peculiar uneasiness rather than true anxiety on free-time days. The lack of structure to the days opens them up to choice, and choices without a guiding pattern may lead either to temptation or to reflection, which then leads to a feeling of not knowing how to act, of existing without purpose.[31]

Most people, however, prefer Fridays to Mondays and usually seem to resent having to work during the weekend, despite the considerable benefits and advantages it often entails—higher pay, lighter traffic to and from work, and a far more relaxed working atmosphere (less supervision, a rather informal dress code, and so on). That is why hospital staff, for example, who can easily swap a Wednesday shift or night on call for a Tuesday one, can hardly ever do that with a Saturday shift or night on call, and why "fair" hospital administrators do their very best to avoid scheduling them for work two weekends in a row.[32] It has also been demonstrated that the quality of one's work is somewhat lower on weekends than on ordinary weekdays.[33]

Most of us seem to feel that working during the weekend is essentially doing something we are not supposed to be doing. Furthermore, it also involves making an actual sacrifice. Consider, for example, a single woman who, "plans to work at home on Sunday, but she still thinks of Sundays as family days and being alone then is like being a dateless teen-ager on Saturday night."[34] The price one pays for being out of phase with the conventional weekly work/rest cycle is essentially a social one, namely depriving oneself of the opportunity to interact intensively with those whom one can see mainly during the weekend. That is why students, for example, often find it much easier to study during the weekend when their spouses, roommates, or friends are also busy studying! (For precisely the same reason, incidentally, having our weekly time off on days when our family or friends have to work robs us of much of the pleasure of being off work and often involves some sense of waste, not to mention guilt. Those who work on Sundays and instead take Tuesdays off, for example, still get together with their friends on Saturday nights more than on Monday nights and are often quite at a loss about how they should spend Tuesdays, when most of their immediate social environment is at work.)

The quality of social interaction varies considerably between ordinary weekdays and weekends. The weekend is generally perceived as a relatively exclusive, "private" time period that is usually spent with family, lovers, and friends. (That is why nonintimates usually sound somewhat apologetic when they call us during the weekend.) Given the close relationship between exclusivity and intimacy,[35] weekend social encounters obviously tend to "mean" much more than weekday ones, which accounts for the differential symbolic significance of Saturday night and Wednesday night dates. (After all, even the prostitute portrayed by Melina Mercouri in Jules Dassin's film *Never on Sunday* would restrict her weekend sexual contacts to "nonprofessional" ones!) The link between exclusivity and intimacy also explains why interpersonal barriers are much more permeable and tend to crumble much more easily and rapidly during the weekend.[36] Thus, for example, when they get together during the weekend, even people who normally encounter one another only on businesslike occasions nevertheless interact with each other in a relatively informal manner. Office weekend picnics, usually characterized by a rather casual atmosphere and a relatively informal dress code, are a perfect case in point.

Just as one is not expected to be working during the weekend, one is also "supposed" to date, go out, and generally "have fun" on weekend (that is, Friday and especially Saturday) nights. That makes adolescents and single adults feel particularly embarrassed to be seen in public by themselves (or, still worse for adolescents, with their parents) on those nights.[37] More sadly, it is also on those nights that they tend to feel the loneliest when staying at home by themselves.[38] That is the general message one gets from the popular song with the suggestive title "Saturday Night Is the Loneliest Night of the Week."[39]

The Pulsating Week

Having examined the cultural "qualities" of four of the seven days of the week (Friday, Saturday, Sunday, and Monday), we have ignored the other three. Since most people regard the weekend as the principal "temporal milestone" within the weekly cycle, whatever "qualities" Tuesday, Wednesday, and Thursday might have, they are generally experienced not quite as intensively as the other four days. In order to understand why, we should briefly reexamine the Jewish and astrological weekly cycles. Their superficial resemblance to one another notwithstanding, these two seven-day cycles were of entirely different origins. Thus, they must have also been experienced in rather strikingly different ways.

Having developed within the context of a polytheistic cosmological system, the astrological week essentially consisted of seven days, each of which was assigned to a particular planet-deity. These seven days carried relatively similar "weights," since each of them had its own distinctive cultural significance, quite independently of the others. The Jewish week, by contrast, evolved within the context of a monotheistic cosmological system and, as such, has always entailed the consecration of only one of its seven days to God. Having essentially derived from the periodic observance of the Sabbath, this week revolves entirely around it. As "the pivot on which the entire [Jewish] week turns, the day from which all others are measured,"[40] the Sabbath alone is regarded by Jews as having its own distinctive characteristics, quite independently of the other six days. As for the latter, they are traditionally divided into the three that follow the Sabbath (Sunday, Monday, and Tuesday) and the three that precede it (Wednesday, Thursday, and Friday),[41] so that they are basically distinguished from one another only in terms of their temporal location vis-à-vis that holy day. Friday, to take one obvious example, derives its entire cultural significance in Judaism from the fact that it is the day of preparation for the Sabbath.

This fundamental difference between the ways these two 7-day cycles were experienced is also evident from the two strikingly different systems used for designating their days. As each one of the seven days of the astrological week had its own distinctive cultural significance, it also had a distinctive name that indicated its association with a particular planet-deity. Not so with the six weekdays of the Jewish week, which are essentially "anonymous."[42] These days were designated, from the very start, by ordinal numbers which indicated their temporal distance from the preceding Sabbath—"first in the Sabbath" (or simply "first day"), "second in the Sabbath," and so on.[43] This serves to remind us that their entire cultural significance in Judaism derives from their temporal location relative to the holy day. As pointed out earlier, the unusual status of the Sabbath is also evident from the fact that, back in antiquity, the same word used to denote both the Sabbath and the entire weekly cycle as a whole.

This latter Jewish practice has been preserved to this day in Armenian (where Shapat' denotes both Saturday and the week as a whole) and, in a slightly modified form, in Romany and Serbo-Croatian as well (where Koóroki and Nedelja, respectively, denote both Sunday and the week as a whole). Neither is the practice of designating days of the week by ordinal or cardinal numbers that reflect their temporal distance from some pivotal day exclusively Jewish by any means. We have already seen, for example, various West African tribes that designate days as "market day is tomorrow," "second day of the market," and so on. Moreover, in ancient Rome, of the eight days of the market week, seven were designated in terms of their temporal distance from the eighth one—VIII Nundinas ("the eighth day before the coming market day"), VII Nundinas, VI Nundinas, V Nundinas, IIII Nundinas, III Nundinas, and Pridie Nundinas ("the day immediately preceding the market day").[44] The Jewish practice of designating days of the week in terms of their temporal distance from the preceding Sabbath has also been preserved—although, often, only in part or with some modifications—in languages as diverse as Persian, Arabic, Greek, Portuguese, Icelandic, Estonian, Lithuanian, Latvian, Russian, Polish, Slovak, Czech, Serbo-Croatian, Slovene, Hungarian, Romany, and Armenian. Finally, there are instances where the day immediately juxtaposed to the pivotal day is named relative to it. The Greeks, for example, still call Friday, the day immediately preceding the Sabbath, Paraskeuē, which means "preparation." Similarly, as the day following the Sunday rest, Monday is known as "the day after nonwork" in Polish (Poniedziałek), Russian (Ponedel'nik), Serbo-Croatian (Ponedeljak), Slovene (Ponedeljek), Slovak (Pondelok), and Czech (Pondělí).

In all the above instances, the week seems to be experienced as a sort of wave that "peaks" on some pivotal day. [Regarding the topographic imagery of peaking, note that the Basques indeed call Sunday "high" (Igandea).] The Sabbath, the Lord's Day, the market day, and the weekend are all classic examples of such *"peak days."* The Judeo-Christian week and all market weeks are fundamentally different from the astrological seven-day week as well as from most of the weeks constituting the divinatory calendars of Central America and Indonesia. Their most distinctive characteristic is the fact that they include only one day (or possibly two consecutive ones, as in the case of the modern weekend) that is culturally marked as pivotal, whereas all the other days of the cycle derive their entire cultural significance from their temporal location relative to it. In fact, had it not been for their differential temporal location relative to the "peak days," those other days would have probably been experienced as absolutely interchangeable with one another. In other words, had it not been for the "peak day" around which the entire weekly cycle seems to revolve, none of them would have even existed as days with any cultural significance!

The contrast between the astrological and Jewish versions of the seven-day week is quite analogous to the contrast between a musical passage with no clear phrasing and one where the first quarter note within each bar is distinctively accented. (Consider also the difference between eating a different kind of meat on each day of the week and eating meat on Mondays only.) As becomes quite evident from this musical analogy, only in the Jewish version do we encounter the phenomenon to which I have referred throughout this book as a seven-day *"beat."* The experience of beat is essentially a sensation of a throbbing pulsation, which is why I shall henceforth refer to weekly cycles with "peak days" as *"pulsating weeks."*

Of the two original versions of the seven-day week, it is clearly the pulsating, Jewish one that has had the most profound impact on the temporal organization of modern life. The two other major monotheistic systems that have sprung out of Judaism followed in its footsteps and also adopted the pattern of consecrating only one of a series of seven days. As a result, the modifications of the identity of the original Sabbath by both Christianity and Islam as well as its more recent transformation into the essentially secularized weekend notwithstanding, the sort of "weekly" life most of us lead today basically derives from the original Jewish model of a seven-day pulsating cycle. None of the distinctive "qualities" of Friday and Monday, for example, would have been possible had we derived our notion of the week from the astrological model of a seven-day nonpulsating cycle.

Note, in this regard, that, colloquially, the words "week" and "day" often refer only to the five weekdays, thus excluding the two weekend days, as when someone inquires whether we should get together "during the week" or on the weekend, or when stores advertise their regular "daily" and weekend hours separately. In other words, of the seven days that constitute the weekly cycle, only five are actually counted as "days" that are part of the "week." The reason for that is that the other two are not considered "ordinary" days:

" 'The day' . . . refers, of course, only to the five days Monday through Friday: Saturday and Sunday are not counted days in the ordinary sense."[45]

Ironically, the key to a full understanding of the pulsating week as a phenomenon may actually be found in the etymology of the English word "week" (which, in its old Gothic form *wiko,* was first used as early as the fourth century[46]). The Latin word *vicis,* from which it most probably derived (along with such linguistic cousins as the Dutch *week,* the Finnish *viikko,* the Icelandic *vika,* the Swedish *vecka,* the German *woche,* the Norwegian *uke,* and the Danish *uge*), was associated with such notions as movement, change, turnabout, and alternation. The intimate relation between the pulsating week and alternation seems to provide us with the best account of the way in which we experience this phenomenon.

According to Emile Durkheim, any calendrical interruption of the otherwise continuous flow of time is a product of a fundamental human intellectual need to establish differentiation, as well as alternate, among various existential domains: "it was probably the necessity of this alternation which led men to introduce into the continuity and homogeneity of duration, certain distinctions and differentiations which it does not naturally have."[47] The most significant "break of continuity" identified by Durkheim was that between the sacred and the profane domains[48] and, thus, also between sacred and profane time. The insertion of sacred time intervals into the continuum of profane time substantiates the discontinuity between the sacred and profane domains, and nowhere is this more evident than in the weekly observance of the Sabbath.[49] The ancient Talmudic ruling that travelers who lose count of the days of the week should nevertheless keep observing the Sabbath every seventh day despite the likelihood of its being the "wrong" day makes it quite clear that at the very heart of the institution of the Sabbath lies the periodic alternation between the sacred and the profane along a 6-1 pattern. This structural feature is far more central to Judaism than the actual temporal location of the sacred within historical time.

Durkheim's theory accounts not only for the Jewish experience of the Sabbath, but also for the way most non-Jews, as well as those who are not even particularly religious, experience the modern weekly cycle, which, after all, essentially revolves around a secularized weekend. Durkheim never restricted his notion of "the sacred" to the relatively narrow domain of institutionalized religion and actually referred with it quite broadly to anything that is collectively regarded and treated as "out of the ordinary." Thus, for him, the experience of time essentially involved periodic alternation between the ordinary and the most generally, albeit collectively, defined as "extraordinary." In fact, even his student Henri Hubert, who preceded Durkheim in pioneering the investigation of the rhythmic nature of social life, claimed that all social rhythms are basically products of the interaction between any "critical" (yet not necessarily religiously significant) days and the intervals among them.[50] However, it is Edmund Leach who has probably gone the farthest in explicitly reformulating Durkheim's theory of time in this direction, claiming the primacy of our experience of temporality, "as something discontinuous, a repetition of repeated reversal, a sequence of oscillations between polar opposites."[51] According to Leach,

> with a *pendulum view of time*, the sequence of things is discontinuous; time is a succession of alternations and full stops. Intervals are distinguished, not as the sequential markings on a tape measure, but as repeated opposites . . . the notion that time is a *"discontinuity of repeated contrasts"* is probably the most elementary and primitive of all ways of regarding time.[52]

We can thus view the pulsating week as a *cycle of periodic alternation between ordinary and extraordinary days.* It is the regular pendulumlike pulsation between the "on" and "off" phases of such a cycle that underlies our very experience of a seven-day "beat." The essence of the experience of the pulsating week is the fundamental cultural binary contrast between the extraordinary and the ordinary. The former is symbolically represented by "critical," "peak" days such as the Sabbath, the Lord's Day, the market day, or the weekend, as well as by such events as the Baha'i Nineteen-Day Feast or the festival dedicated to the patron deity of each Aztec "solar" twenty-day week. The latter is symbolically represented by the ordinary so-called "weekdays." Obviously, had there been no contrast between these two types of days, we would not have had a pulsating week at all.

That would have been the case either if all days were "ordinary" and culturally "unmarked," as in cultures that did not develop a weekly cycle, or if they were all equally "marked," as in nonpulsating cycles such as the astrological week and the weeks of the Central American and Indonesian divinatory calendars. However, the main cultural significance of our weekly "peak days" is the fact that they are not ordinary days. That is why, on many wall calendars, they are printed in red rather than in the ordinary black. At the same time, a concept such as "weekdays" could have evolved only within the context of such a contrast—the word has been used at least since 1477 in contrast to the market day and at least since 1546 specifically in contrast to Sunday.

The rather abstract conceptual contrast between the ordinary and the extraordinary is actually substantiated through our own behavior. It is a "message" that is conveyed quite redundantly through a variety of cultural "codes" that are designed to guarantee that we shall not miss it. Hence its ubiquitous manifestation in so many different aspects of our life. Consider, for example, the extreme, albeit instructive, case of Jewish tradition, which requires believers to walk, sit down, wash themselves, and even move things in a distinctive manner, as well as to avoid uttering any profane words and even thinking about worrisome matters, during the Sabbath.[53] All that is supposed to guarantee that they never confuse the extraordinary sacred with the ordinary profane.

The extraordinariness of certain days of the week can also be "encoded" in distinctive eating patterns. Friday, for example, still means "fast day" in both Icelandic (Föstudagur) and Gaelic (Di-h-aoine). The abstract conceptual contrast between the ordinary and the extraordinary can thus be manifested in the somewhat "special" nature of "peak day" meals, which often follow a "pattern which is an inversion of the mundane week."[54] Consider again the case of Jewish tradition, which requires that food consumed on the Sabbath be different from food consumed on regular weekdays.[55] The fundamental conceptual contrast between the sacred and the profane is thus substantiated by such contrasts as the one between the white Sabbath *hallah* bread and the dark rye bread of regular weekday meals, or that between the costly fish of the Sabbath and the salt herring of regular weekdays.[56]

An even more conspicuous way of "encoding" the special "qualities" of certain days of the week is through associating them regularly with distinctive types of clothing.[57] Such association is quite conventionalized, as one can learn from a set of recent Evan-Picone ads showing the same model wearing different sets of clothes, each one corresponding to, and even captioned by the name of, one particular day of the week. Sunday, for example, is represented by a formal attire in which one would attend church or family reunions, Saturday by sportive clothing usually associated with informal occasions, and both Friday and Monday (quite interchangeable, as working days) by "business" clothes. This seems to indicate that the fundamental cultural contrast between the ordinary and the extraordinary is often substantiated through a contrast between our outward appearance on regular weekdays and on weekly "peak days."

Jewish tradition, for example, puts much emphasis on the establishment of a distinctive "Sabbath look," and an entire chapter of the *Mishnah* is devoted to detailed instructions regarding one's outward appearance on Saturday.[58] Likewise, after indicating explicitly that "thy Sabbath garments should not be like thy weekday garments," the Talmud proceeds to specify that, if one cannot afford at least two distinct sets of garments, so that one can be used exclusively on the Sabbath, one should at least wear one's garments "somewhat differently" on the holy day![59] Thus, to this day, in traditional Jewish neighborhoods, one can easily recognize Saturday by men's Sabbath caps and caftans as well as by women's special jewelry.[60] Following in the footsteps of Judaism, Christianity has taken a similar stance with regard to the Lord's Day. (The special preparation toward a distinctive Sunday appearance is evident even from the fact that, in most North European languages, Saturday is still called a day of washing and laundry—Laugardagur in Icelandic, Lördag in Swedish, Lørdag in Norwegian and Danish, Lauodag in Lapp, Laupäev in Estonian, and Lauantai in Finnish.)[61] Yet the very same spirit has essentially been preserved even in the secular world of work. The very same individual who always wears well-ironed suits or high-heeled shoes from Monday through Friday is likely to drop that "professional" look and adopt a "casual" (or even "sloppy") one on weekends, wearing sneakers and dungarees and possibly even skipping shaving or putting on makeup. Furthermore, when people occasionally come during the weekend to a workplace that is usually open only between Mondays and Fridays, one can still easily tell that it is a weekend day by the mere absence of men's ties and women's nylon stockings.

There is no basic difference between a Jew's Sabbath caftan, a Christian's "Sunday best" attire, and a white-collar worker's weekend "workclothes" (e.g., blue jeans). Nor is the common wardrobe of French miners, which consists of one outfit for work and another one for Sundays,[62] fundamentally different from that of Orthodox Jews, where storage space is often divided between weekday and Sabbath clothes.[63] All are manifestations of the symbolic use of clothing for substantiating conceptual contrasts between abstract categories such as the sacred and the profane, work and leisure, public and private, and so on.

The difference between the way we look on Monday and on Sunday is significantly greater than the difference between the way we look on Monday and on Tuesday. That, of course, is also true of the food we eat, the material we read, and even the people with whom we socialize. Ordinary "weekdays" are usually "marked" much less distinctively than "peak days," so that, physiognomically speaking, they are often quite interchangeable. Hence the common depictions of days such as Tuesday and Thursday as "a day without distinction, an ordinary weekday"[64] or "the neither here nor there part of the week."[65]

Thus, the fundamental contrast between weekly "peak days" and ordinary weekdays is and needs to be substantiated through actual behavioral patterns. After all, if we were to work, dress "professionally," and eat ordinary meals even on Saturday and Sunday, these days would obviously "feel" just like any other day of the week. Whether we substantiate this contrast by eating only *hallah* bread or reading only "nonserious," "pleasure" material, however, is quite beside the point. The pulsating week is a cycle of periodic alternation between opposites, and whether its "peak" is ritually "marked" through prayer, special meals, or family get-togethers is quite irrelevant. The only significant sensation that needs to be experienced is that of a binary contrast between the ordinary and the extraordinary, regardless of what the actual substance of that contrast might be. As one can tell from a cultural "message" such as the Michelob beer commercial that admonishes the listener to "Put a little weekend in your week," the essence of weekly "peak days" is the fact that they are the precise opposite of ordinary weekdays.

In establishing a contrast between the extraordinary and the ordinary, the pulsating week also introduces some discontinuity between them. Thus, for example, structuring our forty-hour workweek so that we work eight hours a day for five days and then rest on the other two—rather than, say, working only five or six hours a day seven days a week—obviously allows us to experience for two full days the interruption of the continuous routine of working. In regularly disrupting the otherwise continuous flow of our life, the pulsating week also adds more dimensions to it. Both those who work seven days a week and those who do not work at all often commit themselves, either willingly or out of necessity, to a rather unidimensional mode of existence and may miss such experiences as diversity and transition. That, of course, is also true of both those who party every night of the week and those who never attend parties at all. In other words, whether the discontinuity that the pulsating week establishes is one between the sacred and the profane, consumption and production, active socializing and isolation, freedom and obligation, the domestic and the public, or spontaneity and routine is rather insignificant. From an experiential standpoint, the most distinctive feature of this cycle is the fact that it helps to introduce discontinuity into our life and thus promote its multidimensionality. After all, in providing us with the opportunity to both work and rest, save and spend, diet and feast, and stay at home and go out, the pulsating week definitely helps us to experience the world around us as both ordinary and extraordinary!

Discontinuity: Boundaries and Beginnings

The experience of time as a discontinuous dimension is a distinctively human experience that cannot be explained by physics. To quote Pitirim Sorokin and Robert K. Merton,

> we cannot carry over into social time the characteristics of continuity which is postulated in the Newtonian conception of astronomical time . . . calculations of time are essentially discontinuous. The natural year may be continuous, but that of the calendar has both a beginning and an end, which are frequently marked by temporal hiatus.[66]

Essentially defying its continuity, the experience of time as a discontinuous dimension accounts for the fact that, on many a wall calendar, the days immediately preceding or following a calendar month that begins or ends in midweek are often left blank as if they did not really exist. It is also responsible for institutionalized breaks such as initiation ceremonies, weddings, and birthdays, as well as for the abrupt transitions between different phases of professional careers, and for the habitual association of mornings with a "fresh start" even in organizations that operate around the clock.[67] In short, it underlies our ability to carve out of the temporal continuum segments of duration that can be handled as if they were discrete particles existing in vacuo. Such segmentation of the temporal continuum into discrete quantumlike particles of duration usually involves one of the major cycles that regulate the rhythmic flow of social life, and the week is obviously one of those cycles.

As we all know, within the context of a linear-historical conception of time, the relations between the years 1982 and 1986, for example, can be viewed as absolutely irreversible. We would not expect, however, that such absolute temporal relations in terms of "before" and "after" could also be established within the context of a circular conception of time. Thus, we would expect Monday, for example, to be viewed as falling both four days before and three days after Friday, since both Monday and Friday are essentially experienced as nonhistorical types of days. And yet, the temporal distance between these two days is usually perceived as being four, rather than three, days,[68] which seems to indicate that Monday is generally perceived as preceding, rather than following, Friday. That can obviously be explained only by our experience of time as a discontinuous dimension. After all, once we experience the week as a discrete segment that can be carved out of the temporal continuum, it can very well have both a beginning and an end.

I have already mentioned earlier that, viewed mathematically as a unit of duration, our week is a seven-day interval that can actually begin at any point in time. Furthermore, having already examined the Soviet continuous workweek as well as the West African market week, we know that a single seven-day weekly rhythm can actually involve no less than seven distinct seven-day weekly cycles, each one beginning on a different day of the week and relating to one another like the different voices in a polyphonic, seven-voice fugue.

The week can begin, for example, on Wednesday, as in the case of thirteen-week or fourteen-week academic semesters that begin on a Wednesday and end on a Tuesday. However, it is experienced as beginning on Saturday by most of those who rent beach houses from Saturday through Friday. Consider also the Friday-through-Thursday cycle (evident, for example, from John Lennon and Paul McCartney's song "Lady Madonna")[69] that seems to characterize the experience of both those who get paid every Friday and those in the entertainment industry that view the week as a series of nights beginning with the first night of the weekend. Gearing themselves particularly to weekend audiences, many movie theaters, for example, usually change their films on Fridays. An extreme example is Manhattan's Metro Cinema, which, although changing its films on a daily basis, nevertheless arranges its calendar in rows of weeks running from Friday through Thursday.

Since there are no absolutely unmistakable and inevitable cutoff points along the temporal continuum, the week can obviously begin on any one of its seven days. And yet, there seem to be some clear social conventions regarding where weeks "begin" and "end," which, once again, justifies examining the experience of the week from a sociological, rather than a psychological, perspective.

The original Jewish week, traditionally revolving around the observance of Saturday as "the seventh day,"[70] obviously began on Sunday. It was thus from Sunday that Jews began counting when designating the days of the week by ordinal numbers reflecting their temporal distance from the preceding Sabbath. As we have already seen, Judaism bequeathed this traditional conception of the week as a Sunday-through-Saturday cycle to Christianity, and that was also true later of Islam. That is how Sunday came to be associated in so many European and Middle Eastern languages with "one" (or "first"), Monday with "two" (or "second"), and so on. As we can see in Figure 13, some of these languages also associate Wednesday with "midweek."

The seven-day week is "officially" still perceived as beginning on Sunday and ending on Saturday. That is how it is graphically represented, for example, on nearly all wall calendars that depict the weekly composition of months. It is also along a weekly cycle which begins on Sunday and ends on Saturday that both radio and

FIGURE 13 The Sunday-Through-Saturday Weekly Cycle

LANGUAGE	SUNDAY 1	MONDAY 2	TUESDAY 3	WEDNESDAY 4 (MIDDLE)	THURSDAY 5	FRIDAY 6	SATURDAY 7
Hebrew	Yom rishon	Yom sheni	Yom shelishi	Yom revi'i	Yom hamishi	Yom shishi	
Arabic	Yôm-el-ḥadd	Yôm-el-itnên	Yôm-el-talât	Yôm-el-arba	Yôm-el-ḥamys		
Persian	Yak-shambah	Dú-shambah	Sih-shambah	Chahár-shambah	Panj-shambah		
Armenian		Y-ergushepti	Y-erekshapti	Chorek-shapti	Hinqshapti		
Greek		Deutéra	Trítê	Tetártê	Pémptê		
Portuguese		Segunda feira	Terça feira	Quarta feira	Quinta feira	Sexta feira	
German (dialects)				Mittwoch	Pfinztag		
Icelandic			þriðjudagur	Miðvikudagur	Fimtudagur		
Italian (dialects)				Mezzedima			
Finnish				Keskiviikko			
Lapp				Kaska-wakko			
Estonian				Kesknädal			
Russian				Sreda			
Polish				Środa			
Serbo-Croatian				Sreda			
Slovene				Sreda			
Slovak				Streda			
Czech				Středa			

television stations normally schedule their programs.[71] Finally, it is also in the order "Sunday, Monday, Tuesday . . ." that first-graders are "officially" introduced to the seven-day sequence. Their habitual chanting may later be manifested in nursery rhymes:

> *Yellow stones on Sunday,*
> *Pearls on Monday,*
> *Rubies on Tuesday,*
> *Sapphires on Wednesday,*
> *Garnets or red stones on Thursday,*
> *Emeralds or green stones on Friday,*
> *Diamonds on Saturday.*[72]

The above nursery rhyme, however, actually constitutes an oddity. In fact, almost all the other nursery rhymes in this genre seem to depict the week as a cycle that begins on Monday and ends on Sunday:

> *Solomon Grundy,*
> *Born on Monday,*
> *Christened on Tuesday,*
> *Married on Wednesday,*
> *Took ill on Thursday,*
> *Worse on Friday,*
> *Died on Saturday,*
> *Buried on Sunday,*
> *This is the end*
> *of Solomon Grundy.*

> *Monday alone,*
> *Tuesday together,*
> *Wednesday we walk,*
> *When it's fair weather.*
> *Thursday we kiss,*
> *Friday we cry,*
> *Saturday's hours*
> *Seem almost to fly.*
> *But of all the days in the week*
> *We will call*
> *Sunday, the rest day,*
> *The best of all.*

> *If you sneeze on Monday, you sneeze for danger;*
> *Sneeze on a Tuesday, kiss a stranger;*
> *Sneeze on a Wednesday, sneeze for a letter;*
> *Sneeze on a Thursday, something better;*
> *Sneeze on a Friday, sneeze for sorrow;*
> *Sneeze on a Saturday, see your sweetheart tomorrow;*

Sneeze on a Sunday, your safety seek,
Or the devil will take you for the rest of the week.

Cut [your fingernails] on Monday, you cut them for health;
Cut them on Tuesday, you cut them for wealth;
Cut them on Wednesday, you cut them for news;
Cut them on Thursday, a new pair of shoes;
Cut them on Friday, you cut them for sorrow;
Cut them on Saturday, see your true love tomorrow;
Cut them on Sunday, ill luck will be with you all week.[73]

As we can see in Figure 14, there are also a number of European languages that associate Monday—rather than Sunday—with "one" (or "first"), Tuesday with "two" (or "second"), and so on.[74] Furthermore, even in English, the day that is "officially" supposed to inaugurate the week is nevertheless included in the concept "weekend," which obviously suggests something that is coming to a close!

While the week still "officially" begins on Sunday, most working adults (including those who occasionally work on Sunday) as well as school-attending children nevertheless experience it as a cycle that begins on Monday. (That, of course, is a result of the replacement of the Jewish Sabbath by the Christian Lord's Day as the weekly day of rest in all predominantly Christian societies. Observant Jews as well as most Israelis obviously continue to experience the week as beginning on Sunday, and there is even a Jewish folktale that symbolically portrays the social contrast between Christian and Jew through the calendrical contrast between the practices of starting to count the days of the week from Monday and Sunday, respectively.[75]) Thus, for example, it is along Monday-through-Sunday weekly cycles that all runners organize their training,[76] and nurses may opt for scheduling their weekends on and off duty as unsegmented wholes despite the fact that Saturday and Sunday may actually appear on two separate weekly time sheets.[77] Note, in this regard, that, while nearly all American wall calendars represent months as made up of weeks that begin on Sunday, almost all appointment books and engagement calendars that represent each week on a separate page depict that cycle as beginning on Monday! In fact, as we can see in Figure 15, many desk calendars even include, on the very same page, both modes of representing the week. Whereas, in the part depicting the weekly composition of months, rows of days beginning on Sunday and ending on Saturday clearly represent the "official" conception of the week, "weekly" pages that begin on Monday evidently reflect the way most of us experience this cycle when scheduling our everyday life. (It is probably in order to resolve such inconvenient duality that Auguste Comte, Gaston Armelin, and L. A. Grosclaude, in their calendar reforms which we examined in Chapter

FIGURE 14 The Monday-Through-Sunday Weekly Cycle

LANGUAGE	MONDAY 1	TUESDAY 2	WEDNESDAY 3	THURSDAY 4	FRIDAY 5	SATURDAY 6	SUNDAY 7
Lithuanian	Pirmadienis	Antradienis	Trečiadienis	Ketvirtadienis	Penktadienis	Šeštadienis	Sekmadienis
Latvian	Pirmdiena	Otrdiena	Trešdiena	Ceturtdiena	Piektdiena	Sestdiena	Svētdiena
Estonian	Esmaspäev	Teisipäev	Kolmapäev	Neljapäev			
Russian		Vtornik		Četverg	Pyatnitsa		
Polish		Wtorek		Czwartek	Piątek		
Slovak		Utorok		Štvrtok	Piatok		
Czech		Úterý		Čtvrtek	Pátek		
Serbo-Croatian		Utorak		Četvrtak	Petak		
Slovene		Torek		Četrtek	Petek		
Hungarian*	Hétfő	Kedd					
Basque	Astelehena						
Romany (Gypsy)	Yek dívvus pálla koóroko	Doói dívvuses pálla koóroko	Trin dívvuses pálla koóroko	Stor dívvuses pálla koóroko	Pansh dívvuses pálla koóroko		

* Monday's name in Hungarian literally means "head of the seven." The names of Thursday (Csütörtök) and Friday (Péntek) are phonetic derivations of the corresponding Slavic names.

FIGURE 15 **The Representation of the Week in Desk Calendars**

June

June						
S	M	T	W	T	F	S
						1
2	3	4	5	6	7	8
9	10	11	12	13	14	15
16	17	18	19	20	21	22
23	24	25	26	27	28	29
30						

Monday

3

Tuesday

4

Wednesday

5

Thursday

6

Friday

7

Saturday

8

Sunday

9

4, all opted for designating Monday as the "official" beginning of the week.)

The "Week-At-a-Glance" appointment books many of us use clearly reflect, as well as facilitate, our experience of time as a discontinuous dimension. The fact that each page of these calendars constitutes a discrete context within which we do our planning and scheduling (so that, upon turning a page, we immediately move to a "different" week) definitely reinforces our habitual temporal orientation in terms of isolated particles of time existing in vacuo—"this week," "last week," "next week," "the week of the eighteenth," and so on.

What allows us to carve out of the temporal continuum these discrete seven-day particles of time are the boundaries that supposedly separate them from one another. It is these boundaries that enable us to tell whether two events that are less than seven days apart belong within the "same" week or within "different" weeks. They account for our ability to defy purely chronological distances and refer on Monday to the preceding Friday, which is only three days away, as "last week," yet to the following Saturday, which is five days away, as "this week." As a rule, distances across boundaries are conceptually longer than those within them.[78] Therefore, Monday and the following Sunday, which are six days apart, are usually regarded as belonging within the "same" week and thus also appear on the same page of our weekly appointment books, whereas Sunday and the following Monday, which are only one day apart, are regarded as being located within two "different" weeks and thus also appear on different pages of our appointment books.

The precise location of the boundaries between weekly segments of time turns out to be more than a mere inconsequential academic matter when one has to adhere strictly to such routines as "no more than three times a week" or "at least once a week." In fact, the rigidity and perceived impermeability of interweekly boundaries is very often quite impressive. Thus, for example, in one hospital I studied, nurses were allowed to combine the two weekly days off of two consecutive weeks so as to enjoy an uninterrupted four-day "long weekend" only if the latter would begin on a Friday.[79] As we can see in Figure 16, the perceived impermeability of the boundaries of the hospital week (which officially began on Sunday) obviously made any other four-day "long weekend" virtually impossible, since it would have involved taking off three or four days within the "same" week and thus violated the rule whereby nurses were allowed only two days off every week!

Interweekly boundaries indeed resemble impermeable walls, yet a rather special kind of walls, namely, glass walls, since they are totally invisible. Discontinuity is very often socially constructed,[80]

FIGURE 16   "Long Weekends" and Interweekly Boundaries

F	S	S	M	T	W	T	F	S	S	M	T	W	T	F	S	S

 A day off

☆ An impossible arrangement

and, while the boundaries separating one seven-day particle of time from the next one clearly provide our life with the ubiquitous and unmistakable temporal structure we have explored in Chapter 5, they are nevertheless purely mental creations rather than the actual barriers that exist "out there" which they seem to be. Most of us, for example, can quite easily detect the overall weekly structure underlying a multiple-weekly schedule that involves "averaging." However, looking at the very same schedule without being familiar with the boundary-maintaining weekly cycle[81] may be as perplexing and cognitively disturbing as looking at a calendar that depicts the days of the month in a single row, without breaking them up into weekly segments. (That is why, in such calendars, Sunday is often printed in a distinctive color.) In a way, it is not unlike hearing a continuous flux of sounds such as "sweetgeorgiabrown" without being familiar with the boundary-maintaining morphological building blocks of the English language. Since culture alone sets the boundaries which help us isolate discrete particles of sound in our mind, it is only those who speak English that can carve out of the above continuum of sounds the chain of sensible, meaningful segments "Sweet/Georgia/Brown." Along similar lines, since culture alone is responsible for setting the boundaries that help us isolate segments of time from the rest of the temporal continuum, obviously only those who are familiar with the overall "sociotemporal order"[82] of our environment can actually "see" the boundaries that essentially define discrete particles of time such as the week.

CHAPTER SEVEN

Culture, Not Nature

A Social Convention

It is important to remember that the "qualities" and meanings we associate with the various days of the week are essentially relative. Thus, for example, from the standpoint of someone working on a newspaper that is regularly published on Wednesdays, "Wednesdays seemed to us to be what Saturdays were to the department-store people, plus what Sundays were to the church people."[1] Similarly, throughout a semester in which I teach my most demanding course on Tuesday, that day becomes the pivotal "temporal landmark" around which my entire weekly work/rest rhythm revolves (so that Monday, for example, always "feels" like the day immediately preceding the peak weekly pressure), yet that changes dramatically the very instant I start a new semester in which I teach that course on Friday. As Thomas Schelling has pointed out,

> a weekday is great for going to the dentist *unless the dentist takes the same day off.* Friday is a great day to head for the country, avoiding Saturday traffic, *unless everyone has Friday off.* Tuesday is no good for going to the beach if Wednesday is the day the children have no school; but Tuesday is no good for getting away from the kids if that's the day they don't go to school.[2]

Consider also the "qualities" we normally associate with the weekend days as well as with Friday and Monday. The fact that we are not likely to identify "Monday blues" on Labor Day (which is always observed on Monday) or among those who regularly take Mondays off, for example, seems to indicate that the supposedly "intrinsic" gloominess of Monday is only a function of its temporal location as the day on which the workweek is ordinarily inaugurated. Similarly, the special "qualities" of Friday are only a function of its temporal location as the last day of the workweek (so that, within the context of a four-day workweek that ends on Thursday, we may even expect an expression such as "Thank God it's Thursday"[3]). The very same festive mood and relatively lax atmosphere that one usually associates with Friday, for example, can also be witnessed at many a workplace on the Wednesday immediately preceding Thanksgiving Day as well as on a Tuesday that happens to be the last day of work before the Christmas vacation. Likewise, the very same kind of "pre-weekend" letdown that most workers ordinarily associate with Friday is often experienced on Wednesday evenings by professors who teach on a Monday–Tuesday–Wednesday schedule.

The relativity of the meanings we attach to the various days of the week is even more evident when we examine cultures other than our own. Thus, for example, only those who know that French schools were traditionally closed on Thursdays can appreciate the subtle message conveyed by Balthus in his painting "The Week with Four Thursdays." Similarly, the fact that Americans' attitude toward Sunday and Monday parallels Iranians' attitude toward Friday (the "peak day" of the Mohammedan week) and Saturday, respectively,[4] serves to remind us that the supposedly "intrinsic" qualities attributed to all those days are actually only a function of their temporal location relative to the essentially conventional, and thus variable, "peak" of the weekly cycle. In this regard, I also recall that the very same gloomy mood I now associate with Sunday evenings I used to experience on Saturday evenings when living in Israel, where the workweek begins on Sunday rather than Monday. (Similarly, dateless Friday nights are associated in Israel with the same kind of loneliness Americans usually associate with dateless Saturday nights.)

Not only the meaning of a given 24-hour interval, but even the decision whether it is a Sunday or a Monday, is essentially a matter of social convention alone. When it is Monday afternoon in Western Europe, for example, it is not absolutely clear, from a purely geographical standpoint, whether it is Monday or Sunday night in the Philippines, which, given the spherical shape of the earth, lie both 130° to the east and 150° to the west of Western Europe. Prior to the introduction of the International Date Line, the very same day which the Spaniards, who had reached the East Indies from America (that is, by moving westward), regarded as Wednesday was regarded as Thursday by the Portuguese, the Dutch, and the English, who had arrived there by way of Africa (that is, by moving eastward)! Likewise, in Alaska, the very same day observed by Americans as Sunday was claimed by Russians to be Monday.[5] Obviously, when the International Date Line was finally established in 1884, the decision where it would pass was also a matter of an arbitrary social convention.[6]

The above discrepancy was empirically discovered only in 1522, by the survivors of Ferdinand Magellan's first voyage around the world,[7] yet the problem itself, as well as the conventionality of any solution of it, was theoretically recognized by Jewish scholars as early as the twelfth century.[8] This is hardly surprising, given the traditional Jewish concern with the precise anchoring of the Sabbath within historical time.[9] This concern has manifested itself in an anxiety (which goes back at least 1,500 years[10]) regarding the possibility of losing count of the days of the week, as well as in the attention devoted to devising methods of keeping count of them. In Jewish fiction, for example, we encounter such routines as piling logs on the stove, tying knots in a rope, or arranging one's hair in a varying number of braids, so that the number of logs, knots, or braids always corresponds to the day of the week (one on Sunday, two on Monday, and so on), thus allowing one to easily reckon what day it is by merely counting them.[11] Consider also the possible time-reckoning function of the seven maids who, according to the Talmud, served Queen Esther on the seven days of the week.[12]

The above routines essentially resemble that—practiced by Sunday-observers such as the Saulteaux Indians or the fictional Robinson Crusoe—of making a mark or carving a notch on a regular daily basis, with Sunday being distinguished from the other days by a particularly longer mark or notch.[13] They are also functionally analogous to the ancient Egyptian and Roman routine of moving pegs across a set of holes[14] or the astrological practice of alternating among a set of rings bearing the names of the planets.[15] (These holes or rings would represent the set of the days of the week, so that, by merely glancing at one's ring or at the hole accommodating the peg, one would easily be able to tell what day it was.) These methods also resemble our own practices of crossing off days on the calendar or encircling them with a magnetic ring. All these routines share one basic feature in common—in order to be, effective, they must be performed on an absolutely regular daily basis. After all, if one would not carve a notch every morning upon waking up or add a braid every night upon going to bed, for example, it would be quite impossible to recall with absolute certainty whether that routine was last performed two, three, or four days ago.

As Francis Colson pointed out, reckoning the day of the week is a "form of time-measurement, which if once lost by a single lapse would be lost for ever."[16] After all, unlike the day or the year, the week is not anchored in nature in any way, so that none of the numerous clues provided by nature to those who wish to reckon the season or the time of day (the temperature, the position of the sun in the sky, the color of the leaves, and so on) can ever be of any help to those who have lost count of the days of the week. Thus, while it is quite unlikely that anyone would ever mistake a day for a night or summer for winter, it is not at all uncommon for someone to make gross errors when trying to identify the day of the week.

Given its considerable temporal regularity, our social environment can easily function as a most reliable clock or calendar.[17] Since numerous events in our everyday life are routinely as well as distinctively associated with particular days of the week, each of the latter has acquired a distinctive "physiognomy,"[18] to the point where it can actually be recognized by these events. Thus, for example, even very young children can already recognize weekend days by the distinctively slower pace of their parents' morning activities. As a child, I myself would recognize Saturdays, immediately upon waking up, by the baroque music on the radio, which was distinctively associated with—and, thus, reliably indicative of—Saturday mornings. Church bells, which indicate that it is Sunday, and garbage-collecting trucks, which signal that it is the regular garbage pick-up day, often acquire a similar cognitive function, as do various other morning sights that are routinely as well as distinctively associated with particular days of the week: "On Sunday mornings early the young wife came out, gathered a handful of the most beautiful roses, and put them in a glass of water, which she placed on a side table. 'I see now that it is Sunday,' said the husband as he kissed his little wife."[19] Throughout the day, similar clues keep forcing themselves on our attention. Consider, for example, my friend who can recognize winter Mondays by the temperature of his office, which normally takes a few hours to heat up after the weekend, or another colleague of mine, who once entered the Faculty House around lunchtime, smelled fish, and immediately remarked: "Oh, today is Friday." Note also the following experience of a man whose lover visits him, regularly as well as exclusively, on Tuesdays: "You are a sort of special clock that counts the weeks instead of the hours. . . . Whenever you arrive, whenever I hear the sound of your doorbell ring, I know that a week has passed."[20]

The physiognomic characteristics of particular days of the week often presuppose familiarity with a particular social environment and are too idiosyncratic to apply beyond it. Thus, upon recalling his family's routine of having lunch an hour earlier than usual on Saturdays, Marcel Proust carefully adds that, while that "asymmetry" of Saturday definitely gave it a somewhat distinctive character, it could not have been noticed by anyone who was not a member of his family.[21] Along similar lines, only those who live on a block where garbage is regularly picked up on Thursday mornings might recognize Wednesday evenings by the trash bags on the sidewalk. A few days of the week, however, have such distinctive physiognomies that they can normally be recognized by entire communities, if not societies. Thus, for example, within a traditional Jewish environment, one would normally have very little trouble recognizing Saturday: "Only by putting your head inside the door, they say, just by sniffing the atmosphere of the house, you can tell whether it is Sabbath or weekday."[22] Similarly, within a predominantly Christian environment, Sunday can be easily recognized by the way people are dressed as well as by the dramatic change in human density (as demonstrated, for example, by Kevin Lynch, through a photographic contrast between the very same street corner on a regular weekday and on a Sunday[23]). Also, throughout the United States, it would be very hard not to recognize Saturday by the crowded supermarkets, laundromats, and department stores, as well as by the long lines in front of movie theaters.

One Friday night several years ago, my wife and I worried that our little daughter might wake us up early the following morning, assuming that it is a regular school day. Looking for an effective way of communicating to a child who still could not read that it was a weekend day (that is, that it was all right for us to sleep late), I decided to leave her weekly allowance, which she regularly received on weekends, rather conspicuously on her bed table. As it turned out, I was quite successful in conveying my message through that rather unorthodox semiotic code. On the following morning our daughter indeed did not wake us up, later explaining that she knew it was a nonschool day the very moment she saw the money!

Chaim Nachman Bialik's humorous short story "The Short Friday"[24] revolves around a similar theme. Late Friday night, long past the time beyond which traveling is strictly prohibited for observant Jews, a rabbi arrives at a Jewish inn. The utter inconceivability of a rabbi traveling during the Sabbath leads the innkeeper, who finds him sleeping there the following morning, to the erroneous conclusion that it must be an ordinary weekday rather than Saturday. (I once witnessed a colleague of mine make a similar sort of error. Upon seeing at our office on Thursday a part-time typist who used to work there on Mondays, Wednesdays, and Fridays only, he convinced himself that it was actually Wednesday.) Worrying that the rabbi would find out that he has almost observed the Sabbath on the "wrong" day, he then decides to transform the inn so that it would look as if it were indeed an ordinary weekday. Given the general "visibility" of the Sabbath within traditional Jewish environments, he removes all cultural items which alone would indicate that it is Saturday: "To begin with, he put away the brass candlesticks, the remains of the Sabbath meal, and the white tablecloth. . . . And straightway the whole appearance of the house was transformed. Sabbath departed and weekday arrived."[25] The innkeeper ushers in the weekday by deliberately performing activities which, in a traditional Jewish environment, could never possibly take place during the Sabbath. He thus starts up a fire, stokes up the samovar with fuel so that it will hum, has his daughter peel potatoes and his hostler chop wood and fix things so that the sounds of hammer and ax will be heard, and winds his philacteries around his arm while repeating the morning prayers to their ordinary weekday tune. When the rabbi finally wakes up, no verbal interchange between them is even necessary. From merely glancing at his surroundings the rabbi can tell that it is definitely an ordinary weekday and not Saturday. The evidence is so overwhelming that, despite recalling having gone to sleep on Friday night, he convinces himself that he slept through Saturday and that it is already Sunday, ironically making the same sort of mistake made earlier by the innkeeper.

Such errors, and the unrecognizability of the days of the week in general, are obviously most typical of those environments where the weekly "beat" is least salient. Bialik's story takes place in an isolated Jewish enclave surrounded by an overwhelmingly non-Jewish environment.[26] Along similar lines, a Talmudic discussion of the possibility of losing count of the days addresses situations whereby travelers are removed from any contact with their Sabbatarian home environment.[27] Such is also the situation of Queen Esther (a Jewess living in the non-Jewish court of the king of Persia) and Robinson Crusoe (a sailor living alone on an island), as well as of anthropologist Irving Hallowell, upon spending a summer among the Pekangikum Indians: "the days of the week became meaningless, since, in two settlements, there were no missionaries and hence no Sunday observance or other activities that differentiated one day from another."[28]

All this indicates that the distinctive physiognomy of the various days of the week may not be as intrinsic to them as it appears to be, and that, if the entire weekly "beat" were to be eliminated, even the particularly "colorful" days would have been "bleached." To quote Pitirim Sorokin, "if there were neither the names of the days nor weeks, we would be liable to be lost in an endless series of days— as gray as fog—and confuse one day with another."[29] Yet errors in identifying the day of the week can and do occur not only on deserted islands or among illiterate tribes, but even within our own society, which is definitely organized in accordance with a weekly "beat." After all, as the experiments of Asher Koriat and others seem to indicate, the recognizability of the various days of the weekly cycle is only a function of their temporal proximity to its "peak," and since the weekend constitutes the most accessible "temporal landmark" within the week, it is usually much easier to recall what day it is on Saturday than on Wednesday.[30] Consequently, the perceived difference between any two days is not necessarily identical to that between two others, and we are much more likely to mistake a Wednesday for a Thursday than a Sunday for a Monday, despite the fact that the temporal distances between each couple of days are chronologically identical. In general, of the seven days of the weekly cycle, only Friday, Saturday, Sunday, and Monday are culturally marked in such a distinctive manner that they are not likely to be confused with one another. Tuesday, Wednesday, and Thursday, on the other hand, usually lack any distinctive physiognomic characteristics (except in a few environments, such as school, where each one of the five weekdays is distinctively associated with a unique activity pattern). Not being as easily recognizable as the former four days of the week, they are quite often mistaken for one another.

However, there are a number of instances when even a Sunday and a Wednesday may be mistaken for one another. After all, it is essentially the regular seven-day work/rest "beat"—and in "work" I also include study—that provides modern life with its distinctive weekly structure. Thus, whereas traditionally it was the absence of a Saturday or Sunday observance that used to evoke the likelihood of losing count of the days of the week, today it is the absence of regular contact with a weekly structured work environment that is normally responsible for making the weekly rhythm less salient. Being outside the work force is thus a major form of being literally out of phase with the rest of society, since, with the main exception of the particularly observant, who attend weekly religious services, those who do not work regularly generally "feel" the seven-day "beat" somewhat less than those who do. The elderly retired, the unemployed,[31] and students who drop out of college and "do nothing" for several months, for example, essentially inhabit environments that are not structured along the weekly cycle, and therefore lack the social clues that normally help us identify a given day as a Friday or a Monday. Within such environments, the various days of the week (including the weekend days) usually lose their ordinary social significance and, thus, also much of their distinctive physiognomy. Being basically quite interchangeable with one another in terms of one's daily routines, days such as Sunday and Wednesday, for example, often "look" and "feel" quite the same and can be easily mistaken for one another.

Yet even for those who do work on a regular basis the normal work/rest rhythm is sometimes interrupted temporarily, making the identity of the various days of the week much less obvious. Vacations and sickness spells are perfect examples of periods when we usually lose all track of the weekly cycle. On the first few days we can still recall what day it is by clinging to the memory of what day yesterday was. However, as time goes on, we normally lose count of the days and have no idea (or concern, for that matter) "where" in the week we are. During such periods all days essentially "look" and "feel" the same, and it is not uncommon for a Friday, for example, to be mistaken for a Monday. (A couple of years ago I used to pick up my daughter every Wednesday not from the bus stop, as I would on any other weekday, but from her school. Interestingly enough, the only time I forgot that Wednesday routine during that entire year was immediately following my return from a brief vacation "away from civilization" in an environment where the weekly cycle was of no relevance to me whatsoever.)

Even more striking are holidays, when even a single weekday away from work interrupts the conventional weekly rhythm of working for five days and then resting for two, and often involves considerable mix-ups. A typical case in point is Labor Day, which, being observed on Mondays, always involves a continuous three-day "weekend." Not only does that day, as the last day of the "weekend," usually "feel" like Sunday; the following Tuesday and Wednesday, which are the day inaugurating the workweek and the day immediately following the day of return to work, often "feel" like Monday and Tuesday, respectively. (For similar reasons, when visiting Israel, where the workweek normally begins on Sunday, one often mistakes Monday for Tuesday, which is the day one ordinarily associates with being the second day of the workweek.) Consider also Thanksgiving Day, which essentially involves a continuous four-day "weekend." As the last day of work, the Wednesday immediately preceding the holiday often "feels" like Friday, whereas Thursday and Friday, being the first two days off work, often "feel" like Saturday and Sunday, respectively.

The essence of the experience of the pulsating week is the periodic discontinuity established between ordinary and extraordinary days. "Colorful" days with a distinctive physiognomy are thus going to be "bleached" whenever the regularity of this alternation between the ordinary and the extraordinary is interrupted. The interruption of the regular weekly work/rest rhythm should thus affect equally those who are away from a work environment and those who are in it continuously. And, indeed, Tuesdays and Saturdays all "look" and "feel" the same not only to those who do not work, but also to those who work seven days a week and do not rest on a regular weekly basis—new mothers, soldiers during war, students preparing for final examinations, and so on.

Finally, note that one's immediate social environment is as crucial as one's own activity cycle for "anchoring" one within the weekly cycle. Thus, Sunday, for example, may feel like an ordinary weekday to someone whose spouse or roommate goes to work.

Reification

Despite all this, the ubiquitous presence of the week in our life and the rigid manner in which it regulates so many of our routine activities often lead us to regard the meanings that we ourselves have attached to the various days of this cycle as if they were intrinsic to them. In other words, the regular, fixed association betwen particular days of the week and particular activity patterns often imbues the "qualities" of these days with an aura of inevitability. Thus, for example, for the average American five-year-old, Sunday is already necessarily associated with being away from kindergarten. Only much later in life, if such a child will encounter a society that is not predominantly Christian, will he or she realize for the first time that the connection between Sunday and being away from school or work is not that inevitable.

The perceived inevitability of the "quality" of the various days of the week is also evident from the worldwide practice of designating them as lucky, unlucky, or even evil,[32] as well as from the herbalist lore regarding the particular days of the week on which particular herbs ought to be picked up in order to be effective.[33] Consider also, in this regard, the perceived connectedness between people's character and the day of the week on which they were born, as manifested, for example, in the following famous nursery rhyme:

Monday's Child is fair of face
Tuesday's Child is full of grace
Wednesday's Child is full of woe
Thursday's Child has far to go
Friday's Child is loving and giving
Saturday's Child works hard for its living

And a Child that is born on the Sabbath Day
Is fair and wise and good and gay.

This connectedness, which we have already encountered within the context of the Central American divinatory calendar, is also evident from ancient Jewish sources[34] as well as from the traditional West African custom of naming babies in accordance with the day of the week on which they are born.[35]

The conventionality of the week is evident not only from the occasional unrecognizability of the days constituting it and from the relativity of their meanings, but also from the cross-cultural variability in the length of this cycle. Throughout this book, we have encountered three-day, four-day, five-day, six-day, eight-day, nine-day, ten-day, twelve-day, thirteen-day, nineteen-day, and twenty-day weeks, aside from our seven-day cycle. Such an overwhelming variability in the rhythm of activity within a single species obviously cannot be attributed to nature, and is clearly a product of the human interference with the natural order of things. To quote Pitirim Sorokin, "natural and cosmic time does not have, *per se,* any such weekly cut. If one thinks that our seven-day week is derived from the phases of the moon (cosmic factors) and coincides with them, he is entirely mistaken. . . . *The week unit is a social convention.*[36] That is precisely why the week provides the best possible context for studying the distinctively human relation to time. The invention of the week was a major breakthrough in the history of human civilization, since it constituted the first successful attempt ever to establish a major cycle of activity that rests on mathematical regularity alone. The week is the only major rhythmic component of our environment that is essentially artificial and totally oblivious to nature.

And yet, despite its obvious conventional nature, we usually surround the seven-day week by an aura of inevitability, essentially overlooking the fact that it is fundamentally different from other major rhythms of human activity, such as the day and the year, in that it is an artifact created by human beings that is entirely independent of any natural periodicity. As a young boy, I myself believed that all Saturdays are sunny. And I still recall an incident back in fourth grade, when our teacher asked when the desert dunes normally migrate and one of my classmates seriously replied: "On Saturday." Similarly, I knew a boy who used to believe that he must have been born on a weekend, since his mother would have probably been too busy working during the week. Yet children are by no means the only ones to confuse social and natural rhythmicity:

> The seven-day week is so universally accepted and the Sabbath as a
> weekly rest-day is so firmly rooted an institution that hardly anyone

questions their origin. It is generally assumed that the seven-day week ending with a day of rest is part of the scheme of creation, and that like the laws of Nature it has always existed and will continue to exist forever.[37]

Many adults maintain that the seven-day week is an integral—and, thus, inevitable—part of nature. They are usually not ready to accept such counter-evidence as the existence of numerous weekly cycles that are not seven days long. When confronted with the Roman eight-day week or the Soviet six-day week, for example, they usually claim that these are not weeks at all, using the tautological argument that they are not seven days long!

A particular manifestation of the confusion between natural facts (which are entirely independent of any human influence) and what Emile Durkheim identified as "social facts,"[38] mistaking social for natural rhythmicity is a perfect example of reification, which is the denial of the fact that social institutions are created by humans alone:

> Reification is the apprehension of human phenomena as if they were things, that is, in non-human or possibly supra-human terms. . . . reification is the apprehension of the products of human activity *as if* they were something else than human products—such as facts of nature, results of cosmic laws, or manifestations of divine will. Reification implies that man is capable of forgetting his own authorship of the human world.[39]

Through reification, the conventionality of a social institution such as the seven-day week is thus transformed into perceived inevitability.

As Peter Berger has pointed out, it is very often religion that most effectively legitimates the process of reification:

> All institutions possess the character of objectivity and their legitimations, whatever content these may have, must continuously undergird this objectivity. The religious legitimations, however, ground the socially defined reality of the institutions in the ultimate reality of the universe, in reality "as such." The institutions are given a semblance of inevitability, firmness and durability that is analogous to these qualities as ascribed to the gods themselves.[40]

Thus, through the association of the Sabbath observance with God's rest following the six days of the Creation, the week has traditionally been explained as an integral part of a divine schema. Obviously, with the modern, secularized substitution of scientism for theology, Nature has come to replace God as the most compelling explanatory principle, so that the origin of the week is much more likely to be linked today to the lunation than to the Creation. In both cases, however, what is essentially a conventional social institution is nevertheless portrayed as inevitable.

Hence the advantage of cultivating a comparative—historical as well as cross-cultural—perspective on the week, mainly through acquaintance with exotic time-reckoning systems such as the Indonesian, Maya, Baha'i, and French Republican calendars, to name but a few. Such a perspective helps us question the perceived inevitability, and thus unveil the actual conventionality, of such reified social artifacts as the seven-day "beat." It helps us become more aware of the fact that, despite the pervasive—often constraining—presence of this rhythm, it is only us who created it in the first place. To quote a passage from Peter Beagle's novel *The Last Unicorn:*

> When I was alive, I believed—as you do—that time was at least as real and solid as myself, and probably more so. I said "one o'clock" as though I could see it, and "Monday" as though I could find it on the map. . . . Like everyone else, I lived in a house bricked up with seconds and minutes, weekends and New Year's Days, and I never went outside until I died, because there was no other door. Now I know that I could have walked through the walls.[41]

Precisely here lies the main significance of studying the seven-day week. Examining this distinctively human rhythm helps us shed more light on the fundamental, yet still rather murky, difference between natural inevitability and social conventionality. Unveiling this difference, and thus eliminating the common confusion, between the natural and the social temporal order helps illuminate the extent to which social conventions influence the way we normally order our lives. And this awareness of the conventionality of social reality may help us discover the potential flexibility that awaits those who venture beyond the wall of what appears to be an inevitable rigid order.

Notes

Introduction

1. David Kantor and William Lehr, *Inside the Family* (San Francisco: Jossey-Bass, 1975), p. 81.
2. Leo Tolstoy, "The Death of Ivan Ilych," in *The Death of Ivan Ilych and Other Stories* (New York: Signet Classics, 1960), p. 139.
3. See, for example, Peter L. Berger and Thomas Luckmann, *The Social Construction of Reality* (New York: Doubleday Anchor, 1967), p. 28.
4. See, for example, Sholem Asch, *Kiddush Ha-Shem* (Tel-Aviv: Dvir, 1953), p. 13.
5. *Newsweek*, February 8, 1982, p. 46.
6. Daniel Defoe, *The Life and Strange Surprising Adventures of Robinson Crusoe, Mariner* (London: Oxford University Press, 1972), p. 64.
7. Dalton Trumbo, *Johnny Got His Gun* (New York: Bantam, 1970), p. 143.
8. *The Babylonian Talmud* (London: Soncino, 1938): *Shabbath* 69b.
9. Gustav Ichheiser, *Appearances and Realities* (San Francisco: Jossey-Bass, 1970), p. 8.

Chapter One

1. Hutton Webster, *Rest Days* (New York: Macmillan, 1916), pp. 196–206.
2. Daniel 10.2–3.

3. Deuteronomy 16.9; Leviticus 23.15.
4. *The Babylonian Talmud* (London: Soncino, 1938): *Ta'anit* 29b; *Midrash Rabbah* (London: Soncino, 1951): *Genesis* 11.5.
5. Genesis 2.2–3.
6. Exodus 20.8–11. See also Exodus 23.12, 34.21, 35.2–3; Deuteronomy 5.12–14.
7. Exodus 31.15–17.
8. William A. Heidel, *The Day of Yahweh* (New York: The American Historical Association, 1929), pp. 397–441.
9. Johannes Hehn, *Siebenzahl und Sabbat bei den Babyloniern und im Alten Testament* (Leipzig: Hinrichs, 1907), pp. 106–9. See Exodus 16.30, 23.12, 34.21, and Leviticus 23.32.
10. Genesis 50.10; Exodus 22.29, 29.30–37; Leviticus 8.33, 12.2, 13.4–34, 14.8, 15.13–28, 22.27, 23; Numbers 19.11–19; I Samuel 31.13; I Kings 8.65; I Chronicles 10.12.
11. Genesis 4.24, 7.2–3, 21.28–31, 29.20–30, 33.3, 41; Exodus 25.37; Leviticus 4.6–17, 16.14; Numbers 19.4, 28.11; Joshua 6; I Kings 7.17; II Kings 4.35, 5.10; Ezekiel 40.22; Zechariah 4.2; Psalms 12.7; Proverbs 6.31; Job 42.8; Daniel 4.13–29; I Chronicles 15.26.
12. Hehn, *Siebenzahl und Sabbat*, pp. 4–6, 11–20, 34–46; Hildegard and Julius Lewy, "The Origin of the Week and the Oldest West Asiatic Calendar," *Hebrew Union College Annual* 17(1942–43):16–18.
13. J. van Goudoever, *Biblical Calendars* (Leiden: E. J. Brill, 1961), pp. 26–29, 173–74; Lewy and Lewy, "The Origin of the Week"; Julian Morgenstern, "The Calendar of the Book of Jubilees, Its Origin and Its Character," *Vetus Testamentum* 5(1955):45–46, 56–59, 70–71.
14. Morgenstern, ibid., pp. 39, 55.
15. Hehn, *Siebenzahl und Sabbat*, pp. 106–9; S. Langdon, *Babylonian Menologies and the Semitic Calendars* (London: Oxford University Press, 1935), pp. 83–87, 95–96; Webster, *Rest Days*, pp. 223–35.
16. F. K. Ginzel, *Handbuch der Mathematischen und Technischen Chronologie* (Leipzig: J. C. Hinrichs, 1911), Vol. 2, p. 9; Heidel, *The Day of Yahweh*, pp. 363, 436; Morris Jastrow, *Hebrew and Babylonian Traditions* (New York: Scribner's Sons, 1914), p. 164; Langdon, ibid., pp. 85, 89–90; Theophile J. Meek, "The Sabbath in the Old Testament," *Journal of Biblical Literature* 33(1914):209–10.
17. J. Mann, "The Observance of the Sabbath and the Festivals in the First Two Centuries of the Current Era according to Philo, Josephus, the New Testament, and the Rabbinic Sources," *The Jewish Review* 4(1913–14):443.
18. Hehn, *Siebenzahl und Sabbat*, pp. 42, 113.
19. Francis H. Colson, *The Week* (Cambridge, England: Cambridge University Press, 1926), p. 3.
20. Charles Dundas, "Chagga Time Reckoning," *Man* 26(1926):141; A. B. Ellis, *The Yoruba-Speaking Peoples of the Slave Coast of West Africa* (London: Chapman and Hall, 1894), p. 144; Ginzel, *Handbuch*, Vol. 2, pp. 320–25; Benjamin D. Meritt, *The Athenian Year* (Berkeley and Los Angeles: University of California Press, 1961), pp. 38–59; William Skinner,

"Marketing and Social Structure in Rural China," *Journal of Asian Studies* 24(1964–65):3–43, 195–228, 363–99; Webster, *Rest Days*, pp. 154–66, 188–94.

21. Webster, ibid., p. 194.
22. Ibid., p. 46.
23. Eviatar Zerubavel, *Hidden Rhythms* (Chicago and London: University of Chicago Press, 1981), pp. 9–12.
24. Rudolph von Ihering, *The Evolution of the Aryan* (New York: Henry Holt, 1897), p. 117; Max Weber, *Ancient Judaism* (New York: Free Press, 1967), p. 151; Webster, *Rest Days*, p. 102. On the parallel evolution of the standard hour, see Zerubavel, *Hidden Rhythms*, pp. 37–39.
25. Auguste Bouché-Leclercq, *L'Astrologie Grecque* (Brussels: Culture et Civilisation, 1963), pp. 475–76; Georges Daressy, "Une Ancienne Liste des Décans Égyptiens," *Annales du Service des Antiquités de l'Égypte* 1(1900):79–90; Ginzel, *Handbuch*, Vol. 1, p. 165; Margaret A. Murray, *The Splendor that Was Egypt* (New York: Hawthorn, 1963), p. 286; Otto Neugebauer, *The Exact Sciences in Antiquity* (New York: Harper Torchbooks, 1962), pp. 82–86; Flinders Petrie, *Wisdom of the Egyptians* (London: British School of Archaeology in Egypt, 1940), p. 11; Webster, *Rest Days*, p. 191.
26. Meek, "The Sabbath," pp. 209–10.
27. Zerubavel, *Hidden Rhythms*, pp. 105–10.
28. Jastrow, *Hebrew and Babylonian Traditions*, p. 173; Weber, *Ancient Judaism*, p. 150; Webster, *Rest Days*, pp. 253–55.
29. Lewis Mumford, *Technics and Civilization* (New York: Harbinger, 1963), pp. 197–98.
30. Daniel J. Boorstin, *The Discoverers* (New York: Random House, 1983), pp. 12–13.
31. Udo Strutynski, "Germanic Divinities in Weekday Names," *Journal of Indo-European Studies* 3(1975):364, 372–75; Webster, *Rest Days*, p. 172.
32. Alfred L. Kroeber, *Anthropology* (New York: Harcourt, Brace, 1948), p. 485.
33. Bouché-Leclercq, *L'Astrologie Grecque*, pp. 477–78; Hehn, *Siebenzahl und Sabbat*, p. 52; Webster, *Rest Days*, p. 215.
34. Neugebauer, *The Exact Sciences*, p. 187; A. Sachs, "Babylonian Horoscopes," *Journal of Cuneiform Studies* 6(1952):54–57; B. L. van der Waerden, "The Date of Invention of Babylonian Planetary Theory," *Archive for History of Exact Sciences* 5(1968):71.
35. J. P. V. D. Balsdon, *Life and Leisure in Ancient Rome* (New York: McGraw-Hill, 1969), p. 61; Franz Cumont, *Astrology and Religion Among the Greeks and Romans* (New York: Dover, 1960), p. 91; Solomon Gandz, "The Origin of the Planetary Week or the Planetary Week in Hebrew Literature," *Proceedings of the American Academy for Jewish Research* 18(1948–49):216; Hehn, *Siebenzahl und Sabbat*, pp. 51–52; Kroeber, *Anthropology*, pp. 485–86; Neugebauer, ibid., p. 168; Emil Schürer, "Die Siebentägige Woche im Gebrauche der Christlichen Kirche der Ersten Jahrhunderte," *Zeitschrift für die Neutestamentliche Wissenschaft* 6(1905):18; Webster, *Rest Days*, pp. 216–19.

36. Dio Cassius, *Dio's Roman History* (Troy, NY: Pafraets, 1905), Book 37.18–19.
37. F. Boll, "Hebdomas," in *Pauly-Wissowa Real-Encyclopädie der Classischen Altertumswissenschaft* (Stuttgart: J. B. Metzler, 1912), Vol. 7, pp. 2561–65; Cumont, *Astrology and Religion*, p. 91; Gandz, "The Origin of the Planetary Week," p. 220; Otto Neugebauer, *A History of Ancient Mathematical Astronomy* (New York: Springer-Verlag, 1975), p. 690; Webster, *Rest Days*, p. 215.
38. Simon Mitton (ed.), *The Cambridge Encyclopaedia of Astronomy* (New York: Crown, 1977), p. 161; Gerald F. W. Mulders, "Intricacies of the Calendar," *Journal of Calendar Reform* 11(1941):135.
39. Boll, "Hebdomas," p. 2567; Bouché-Leclercq, *L'Astrologie Grecque*, p. 479; Kroeber, *Anthropology*, pp. 485–86; Neugebauer, *The Exact Sciences*, pp. 169–70; Neugebauer, *A History*, p. 691.
40. Colson, *The Week*, pp. 31–32.
41. Dio Cassius, *Dio's Roman History*, Book 37.17–19.
42. For an elegant, geometric representation, see also Bouché-Leclercq, *L'Astrologie Grecque*, p. 482; Julius C. Hare, "On the Names of the Days of the Week," *Philological Museum* 1(1832):7–8.
43. Bouché-Leclercq, ibid., pp. 481–82.
44. Colson, *The Week*, p. 49.
45. Bouché-Leclercq, *L'Astrologie Grecque*, pp. 474–86; Colson, ibid., pp. 46–49; Cumont, *Astrology and Religion*, p. 91; Gandz, "The Origin of the Planetary Week," p. 222; Kroeber, *Anthropology*, pp. 485–86; James G. Macqueen, *Babylon* (New York: Praeger, 1965), p. 211; Webster, *Rest Days*, pp. 216–18.
46. Boll, "Hebdomas," p. 2574; Ginzel, *Handbuch*, Vol. 2, p. 10; Schürer, "Die Siebentägige Woche," p. 19.
47. On that tradition, see Raymond Klibansky, E. Panofsky, and F. Saxl, *Saturn and Melancholy* (London: Nelson, 1964).
48. *The Babylonian Talmud: Arakin* 11b–12a; *Pesahim* 112b; *Shabbath* 129b, 156a–156b; *Ta'anit* 29b; Louis Ginzberg, *The Legends of the Jews* (Philadelphia: The Jewish Publication Society of America, 1968), Vol. 5, p. 405; George G. Ramsay, *The Histories of Tacitus* (London: John Murray, 1915), p. 402 (Book V, Ch. 4); Joshua Trachtenberg, *Jewish Magic and Superstition* (New York: Atheneum, 1970), pp. 47, 252.
49. Gandz, "The Origins of the Planetary Week," p. 216; Ginzel, *Handbuch*, Vol. 2, p. 10; Hehn, *Siebenzahl und Sabbat*, pp. 120–21; Schürer, "Die Siebentägige Woche," p. 19; Webster, *Rest Days*, p. 244.
50. Dio Cassius, *Dio's Roman History*, Book 37.18.
51. Colson, *The Week*, pp. 24–25; Schürer, "Die Siebentägige Woche," pp. 32–33.
52. Clement of Alexandria, *The Stromata* [pp. 299–567 in Vol. 2 of Alexander Roberts and James Donaldson (eds.), *The Anti-Nicene Fathers* (Grand Rapids, MI: W. B. Eerdmans, 1956)], Book VII, Ch. 12; Colson, ibid., pp. 25–27; Justin Martyr, *The First Apology* (Oxford: J. H. and Jas. Parker, 1861), Ch. 67; Webster, *Rest Days*, pp. 219, 268.

53. Atilius Degrassi (ed.), *Inscriptiones Italiae* (Rome: Libreria dello Stato, 1963), Vol. 13, Fasc. 2, p. 309.
54. Balsdon, *Life and Leisure*, p. 63; Boll, "Hebdomas," p. 2574; Colson, *The Week*, p. 32; Degrassi, ibid., p. 305; Gandz, "The Origin of the Planetary Week," pp. 216, 221; Hare, "On the Names," p. 31; Schürer, "Die Siebentägige Woche," pp. 27–28; Webster, *Rest Days*, p. 216.
55. Balsdon, ibid., pp. 62–63; Boll, ibid., p. 2573; Schürer, ibid., pp. 25–26; Webster, ibid., p. 219.
56. Boll, ibid., p. 2573; Colson, *The Week*, pp. 16–17; Gandz, "The Origin of the Planetary Week," p. 216; Ginzel, *Handbuch*, Vol. 3, p. 98; Schürer, ibid., p. 25; Webster, ibid., pp. 244–45.
57. Mircea Eliade, *Cosmos and History* (New York: Harper Torchbooks, 1959), pp. 51–92; Mircea Eliade, *The Sacred and the Profane* (New York: Harcourt, Brace & World, 1959), pp. 68–113; Barry Schwartz, "The Social Context of Commemoration: A Study in Collective Memory," *Social Forces* 61(1982):375–76; Zerubavel, *Hidden Rhythms*, pp. 108–9.
58. Matthew 28.1; Mark 16.2,9; Luke 24.1; John 20.1.
59. *The Epistle of Barnabas* [pp. 137–49 in Vol. 1 of Alexander Roberts and James Donaldson (eds.), *The Anti-Nicene Fathers* (Grand Rapids, MI: W. B. Eerdmans, 1956)], 13.10.
60. Justin Martyr, *The First Apology*, Ch. 67. See also his *Dialogue with Trypho* (pp. 194–270 in Vol. 1 of Roberts and Donaldson, ibid.), Ch. 41.
61. Van Goudoever, *Biblical Calendars*, pp. 166–67, 174–75.
62. Acts 20.7; I Corinthians 16.2. See also *The Teachings of the Twelve Apostles* [pp. 15–25 in James A. Kleist (ed.), *Ancient Christian Writers* (Westminster, MD: Newman, 1948)], Ch. 14.
63. Schürer, "Die Siebentägige Woche," pp. 12–13.
64. Matthew 27.62, 28.1; Mark 15.42, 16.2,9; Luke 23.54, 24.1; John 19.31, 20.1; Acts 20.7; I Corinthians 16.2; Mann, "The Observance," p. 511.
65. Schürer, "Die Siebentägige Woche," pp. 9–12.
66. Colson, *The Week*, pp. 13–16; Josephus, *Flavius Josephus Against Apion* [pp. 607–36 of his *Complete Works* (Grand Rapids, MI: Kregel, 1960)], II.40; Mann, "The Observance," p. 451; Webster, *Rest Days*, p. 266.
67. Eviatar Zerubavel, *Patterns of Time in Hospital Life* (Chicago and London: University of Chicago Press, 1979), pp. 60–83; Zerubavel, *Hidden Rhythms*, pp. 64–81; Eviatar Zerubavel, "Easter and Passover: On Calendars and Group Identity," *American Sociological Review* 47(1982):284–89; Eviatar Zerubavel, "The Standardization of Time: A Sociohistorical Perspective," *American Journal of Sociology* 88(1982):18–19.
68. Mary Douglas, "The Bog Irish," pp. 59–76 in her *Natural Symbols* (New York: Vintage, 1973).
69. Zerubavel, *Hidden Rhythms*, pp. 70–72.
70. Zerubavel, "Easter and Passover."
71. Abraham E. Millgram, *Sabbath—Day of Delight* (Philadelphia: The Jewish Publication Society of America, 1965), p. 365. See also Germano Pàttaro, "The Christian Conception of Time," in *Cultures and Time* (Paris: UNESCO Press, 1976), p. 183.

72. Colossians 2.16. See also Paul Cotton, *From Sabbath to Sunday* (Bethlehem, PA: Times Publishing Co., 1933), pp. 41–45.

73. Ignatius, *Epistle to the Magnesians* [pp. 59–65 in Vol. 1 of Alexander Roberts and James Donaldson (eds.), *The Anti-Nicene Fathers* (Grand Rapids, MI: W. B. Eerdmans, 1956)], Ch. 9.

74. Zerubavel, "Easter and Passover."

75. Cotton, *From Sabbath*, pp. 54–68; Ephraim Isaac, *A New Test–Critical Introduction to Maṣḥafa Berhān* (Leiden: E. J. Brill, 1973); Webster, *Rest Days*, pp. 269–70.

76. Charles J. Hefele, *A History of the Councils of the Church* (Edinburgh: T. & T. Clark, 1896), Vol. 2, p. 316. See also Alexander Philip, *The Calendar* (New York: Macmillan, 1921), p. 31.

77. Colson, *The Week*, pp. 105–7.

78. Justin Martyr, *The First Apology*, Ch. 67; Clement of Alexandria, *The Stromata*, Book VII, Ch. 12.

79. Boll, "Hebdomas," p. 2578; Schürer, "Die Siebentägige Woche," pp. 35–38; Webster, *Rest Days*, p. 220.

80. Samuel G. Barton, "The Quaker Calendar," *Proceedings of the American Philosophical Society* 93(1949):32–39.

81. Colson, *The Week*, pp. 26, 119.

82. Ibid., pp. 108–12.

83. Ibid., p. 113; Strutynski, "Germanic Divinities."

84. Jacob Grimm, *Teutonic Mythology* (New York: Dover, 1966), Vol. 1, p. 191; Webster, *Rest Days*, p. 221.

85. Tertullian, *Apology* [pp. 17–55 in Vol. 3 of Alexander Roberts and James Donaldson (eds.), *The Anti-Nicene Fathers* [Grand Rapids, MI: W. B. Eerdmans, 1951)], Ch. 16.

86. Balsdon, *Life and Leisure*, p. 64; Colson, *The Week*, pp. 93–94; Cotton, *From Sabbath*, pp. 132–45; Franz Cumont, *Textes et Monuments Figurés Relatifs aux Mystères de Mithra* (Brussels: H. Lamertin, 1899), Vol. 1, p. 119; Franz Cumont, *The Mysteries of Mithra* (New York: Dover, 1956), pp. 120–21; Kroeber, *Anthropology*, p. 487; Webster, *Rest Days*, p. 220.

87. Colson, *The Week*, pp. 92–93.

88. J. F. Fleet, "The Use of the Planetary Names of the Days of the Week in India," *Journal of the Royal Asiatic Society* 44 (1912), pp. 1042–43; Ginzel, *Handbuch*, Vol. 1, pp. 339–40; Pandurang V. Kane, *History of Dharmaśāstra* (Poona: Bhandarkar Oriental Research Institute, 1974), pp. 681–82; Neugebauer, *The Exact Sciences*, p. 169; Swamikannu Pillai, *Indian Chronology* (Madras: Grant, 1911), p. 102.

89. Swamikannu Pillai, ibid., p. 85.

90. Fleet, "The Use," pp. 1044–45; Kane, *History*, pp. 679–82; Webster, *Rest Days*, p. 200.

91. Kroeber, *Anthropology*, p. 487; Webster, ibid., p. 201.

92. Webster, ibid., pp. 197–99, 204–5.

93. Al-Qaṣṭallānī, quoted in S. D. Goitein, "The Origin and Nature of the Muslim Friday Worship," *The Muslim World* 49(1959):183.

94. D. S. Margoliouth, quoted in Goitein, ibid., p. 185.

95. Sherrard B. Burnaby, *Elements of the Jewish and Mohammedan Calendars* (London: George Bell, 1901), p. 387.

Chapter Two

1. On this, see also Morgenstern, "The Calendar"; Shmaryahu Talmon, "Divergences in Calendar-Reckoning in Ephraim and Judah," *Vetus Testamentum* 8(1958):48–74; Zerubavel, *Hidden Rhythms*, pp. 70–81, 98–100; Zerubavel, "Easter and Passover."
2. George B. Andrews, "Making the Revolutionary Calendar," *American Historical Review* 36(1931):516; Henri Welschinger, *Les Almanachs de la Révolution* (Paris: Librairie des Bibliophiles, 1884), pp. 2–3.
3. James Guillaume (ed.), *Procès-Verbaux du Comité d'Instruction Publique de la Convention Nationale* (Paris: Imprimerie Nationale, 1894), Vol. 2, pp. 440–51.
4. Guillaume, ibid., Vol. 2, pp. 588, 887.
5. Zerubavel, *Hidden Rhythms*, pp. 82–95.
6. Guillaume, *Procès-Verbaux*, Vol. 2, pp. 440, 445, 696–97, 876.
7. Ibid., pp. 44, 443–44, 701, 881–82.
8. Ibid., pp. 696, 698, 881.
9. Andrews, "Making the Revolutionary Calendar," p. 525.
10. Ibid., p. 516.
11. Zerubavel, *Hidden Rhythms*, pp. 85–87.
12. Pierre Gaxotte, *The French Revolution* (London: Charles Scribner's Sons, 1932), p. 329.
13. Samuel Seabury, *The Theory and Use of the Church Calendar in the Measurement and Distribution of Time* (New York: Pott, Young, & Co., 1872), p. 216; P. W. Wilson, *The Romance of the Calendar* (New York: W. W. Norton, 1937), p. 235.
14. A. Aulard, *The French Revolution* (New York: Russell & Russell, 1965), Vol. 4, p. 102.
15. L'Abbé J. Gallerand, *Les Cultes sous la Terreur en Loir-et-Cher, 1792–1795* (Paris: Grande Imprimerie de Blois, 1928), p. 634; Henri Grégoire, *Histoire de Sectes Religieuses* (Paris: Baudouin Frères, 1828), Vol. 1, p. 240.
16. Benjamin Bois, *Les Fêtes Révolutionnaires à Angers 1793–1799* (Paris: Félix Alcan, 1929), pp. 20–26; Albert Mathiez, *The French Revolution* (New York: Russell & Russell, 1962), p. 410.
17. Bois, ibid., p. 35; Daniel Guérin, *Class Struggle in the First French Republic* (London: Pluto, 1977), p. 250.
18. For the complete list, see Grégoire, *Histoire*, Vol. 1, pp. 177–78.
19. Gallerand, *Les Cultes*, pp. 727, 732.
20. Aulard, *The French Revolution*, Vol. 4, p. 103; Bois, *Les Fêtes*, p. 158.
21. Gallerand, *Les Cultes*, pp. 637–41.
22. Bois, *Les Fêtes*, pp. 61–66, 69–74, 81–103, 112–14.

23. Ibid., pp. 66–69, 75–78, 101–23.

24. A. Aulard, *Christianity and the French Revolution* (New York: Howard Fertig, 1966), pp. 155–56; Martyn Lyons, *France Under the Directory* (Cambridge, England: Cambridge University Press, 1975), p. 112; Albert Mathiez, *La Théophilanthropie et le Culte Décadaire, 1796–1801* (Paris, 1904).

25. Mathiez, ibid., pp. 414–18.

26. Aulard, *The French Revolution*, Vol. 4, pp. 98–101; Bois, *Les Fêtes*, pp. 127, 154, 189, 198–204; Georges Lefebvre, *La Révolution Française* (Paris: Presses Universitaires de France, 1963), p. 500; Lyons, *France Under the Directory*, p. 242; Mathiez, ibid., pp. 419–21; M. J. Sydenham, *The First French Republic, 1792–1804* (London: B. T. Batsford, 1974), pp. 154, 181; Welschinger, *Les Almanachs*, p. 54.

27. Gallerand, *Les Cultes*, p. 637; Grégoire, *Histoire*, Vol. 1, p. 210.

28. Bois, *Les Fêtes*, p. 154.

29. Ibid., pp. 145–50.

30. Aulard, *The French Revolution*, Vol. 4, pp. 62, 101–6; Aulard, *Christianity*, pp. 152–53; Bois, ibid., pp. 127, 193; Gallerand, *Les Cultes*, p. 739; Mathiez, *La Théophilanthropie*, pp. 198, 424–25, 446.

31. Mathiez, ibid., p. 425.

32. Bois, *Les Fêtes*, pp. 206–18; Mathiez, ibid., pp. 461–532.

33. Bois, ibid., pp. 127, 190; Mathiez, ibid., pp. 428–30, 439–44.

34. Mathiez, ibid., p. 64; Constant Pierre, *Les Hymnes et Chansons de la Révolution* (Paris: Imprimerie Nationale, 1904), pp. 603, 651.

35. Henri Lecomte, *Histoire des Théâtres de Paris: Le Théâtre de la Cité, 1792–1807* (Paris: H. Daragon, 1910), pp. 116–17; Henri Welschinger, *Le Théatre de la Révolution, 1789–1799* (Paris: Charavay Frères, 1880), pp. 450–51; Welschinger, *Les Almanachs*, pp. 55–56.

36. Aulard, *The French Revolution*, Vol. 4, p. 97; Mathiez, *The French Revolution*, p. 474.

37. Andrews, "Making the Revolutionary Calendar," p. 528.

38. Aulard, *The French Revolution*, Vol. 4, p. 102; Guérin, *Class Struggle*, p. 148; Lyons, *France Under the Directory*, pp. 89, 113; Mathiez, *La Théophilanthropie*, pp. 469, 532; Welschinger, *Les Almanachs*, p. 56.

39. Grégoire, *Histoire*, Vol. 1, p. 122.

40. Mathiez, *La Théophilanthropie*, pp. 580–81.

41. Bois, *Les Fêtes*, pp. 241–42; Mathiez, ibid., pp. 589, 599.

42. Bois, ibid., p. 227; Georges Lefebvre, *Napoleon—From 18 Brumaire to Tilsit, 1799–1807* (New York: Columbia University Press, 1969), p. 92; Mathiez, ibid., p. 591.

43. Mathiez, ibid., pp. 599–602.

44. Bois, *Les Fêtes*, p. 234; Gallerand, *Les Cultes*, p. 741; Grégoire, *Histoire*, Vol. 1, p. 335; Lyons, *France Under the Directory*, p. 113; Mathiez, ibid., p. 604.

45. Andrews, "Making the Revolutionary Calendar," p. 531.

46. "The Continuous Working Week in Soviet Russia," *International Labor Review* 23(1931):158–59.

47. Katharine Atholl, *The Conscription of a People* (New York: Columbia University Press, 1931), pp. 84–85; "The Continuous Working Week," pp. 159–62.
48. "The Continuous Working Week," p. 157; Susan M. Kingsbury and Mildred Fairchild, *Factory Family and Woman in the Soviet Union* (New York: G. P. Putnam's Sons, 1935), p. 243; "Russian Experiments," *Journal of Calendar Reform* 6(1936):71.
49. "The Continuous Working Week," p. 167.
50. Atholl, *The Conscription*, p. 85; "The Continuous Working Week," p. 168; Elisha M. Friedman, *Russia in Transition* (New York: Viking, 1932), p. 175; Leo Gruliow, "Significant Russian Approval," *Journal of Calendar Reform* 23(1953):103; Albert Parry, "The Soviet Calendar," *Journal of Calendar Reform* 10(1940):63.
51. P. M. Dubner, "Uninterrupted Week and Labor Productivity," *Predpriyatiye* 73(1929), No. 9, p. 51.
52. Gruliow, "Significant Russian Approval," p. 103; Parry, "The Soviet Calendar," pp. 63–64.
53. Atholl, *The Conscription*, p. 84. See also "The Continuous Working Week," p. 173.
54. William H. Chamberlin, *The Soviet Planned Economic Order* (Boston: World Peace Foundation, 1931), pp. 162–63; Erland Echlin, "How All Nations Agree," *Journal of Calendar Reform* 8(1938):27; Calvin B. Hoover, *The Economic Life of Soviet Russia* (New York: Macmillan, 1931), pp. 248–49; Walter Kolarz, *Religion in the Soviet Union* (New York: St. Martin's Press, 1961), p. 31; Lancelot Lawton, "Labour," in P. Malevsky-Malevitch (ed.), *Russia U.S.S.R.* (New York: William Farquhar Payson, 1933), p. 602; "Russian Experiments," p. 69.
55. Gruliow, "Significant Russian Approval," p. 103.
56. Dubner, "Uninterrupted Week," p. 51; Gruliow, ibid., p. 104; *Izvestia*, March 17, 1930; Kingsbury and Fairchild, *Factory Family and Woman*, p. 246; Kolarz, *Religion in the Soviet Union*, p. 30; Parry, "The Soviet Calendar," p. 67; *Trud*, November 3, 1929, p. 4, and November 5, 1929, p. 4.
57. Echlin, "How All Nations Agree," p. 26; Gruliow, "Significant Russian Approval," p. 104; Carleton J. Ketchum, "Russia's Changing Tide," *Journal of Calendar Reform* 13(1943):149; Kingsbury and Fairchild, ibid., pp. 245–46; Parry, ibid., p. 64.
58. Zerubavel, *Patterns of Time*, pp. 60–61; Zerubavel, *Hidden Rhythms*, pp. 64–68, 128. See also Stanford W. Gregory, "A Quantitative Analysis of Temporal Symmetry in Microsocial Relations," *American Sociological Review* 48(1983):129–35; Alfred Schutz, "Making Music Together: A Study in Social Relationship," pp. 159–78 in Vol. 2 of his *Collected Papers* (The Hague: Martinus Nijhoff, 1964).
59. Wilbert E. Moore, *Man, Time, and Society* (New York: John Wiley & Sons, 1963), pp. 121–22; Zerubavel, *Patterns of Time*, pp. 47–49, 60–61; Zerubavel, *Hidden Rhythms*, pp. 68–69.
60. Echlin, "How All Nations Agree," p. 26; Gruliow, "Significant Russian

Approval," p. 103; Ketchum, "Russia's Changing Tide," p. 150; Moore, ibid., p. 122; Parry, "The Soviet Calendar," pp. 64–65.

61. Seymour M. Lipset, M. A. Trow, and J. S. Coleman, *Union Democracy* (Glencoe, IL: Free Press, 1956), pp. 136–38; Zerubavel, *Patterns of Time*, pp. 75–77, 82–83.

62. "The Continuous Working Week," p. 176. See also ibid., pp. 174–75; Hoover, *The Economic Life*, p. 249; Parry, "The Soviet Calendar," p. 65.

63. Lawton, "Labour," p. 602; Moore, *Man, Time, and Society*, p. 122.

64. Nadezhda Krupskaya, "Culture, Daily Life and the Continuous Week," in *Pedagogicheskiye Sochineniya* (Moscow: APN, 1959), Vol. 6, pp. 150–51.

65. Atholl, *The Conscription*, p. 107; *Izvestia*, March 17, 1930.

66. "The Continuous Working Week," p. 170; Friedman, *Russia in Transition*, pp. 260–61; Gruliow, "Significant Russian Approval," pp. 103–4; Leonard E. Hubbard, *Soviet Labour and Industry* (London: Macmillan, 1942), p. 47.

67. "The Continuous Working Week," pp. 169–170; Friedman, ibid., p. 261; *Izvestia*, March 17, 1930; *Trud*, September 24, 1929, p. 4.

68. Zerubavel, *Patterns of Time*, pp. 46–50.

69. Max Weber, *Economy and Society* (Berkeley: University of California Press, 1978), pp. 217–23, 956–63.

70. Zerubavel, *Patterns of Time*, pp. 43–46.

71. Eviatar Zerubavel, "The Bureaucratization of Responsibility: The Case of Informed Consent," *Bulletin of the American Academy of Psychiatry and the Law* 8(1980): 162–63. See also Bibb Latané and John M. Darley, "Bystander 'Apathy,'" *American Scientist* 57(1969):244–68.

72. "The Continuous Working Week," p. 171; *Izvestia*, March 17, 1930.

73. *Krokodil*, July 1931, p. 9.

74. "Stalin on New Economic Problems," *Soviet Union Review* 9(1931):149–50. See also Atholl, *The Conscription*, pp. 173–74; Friedman, *Russia in Transition*, pp. 261–62; Hubbard, *Soviet Labour*, p. 59; N. Zhukov and A. Shkurat, *Seven Hours—1927–1932* (Moscow: Profizdat, 1933), pp. 53–56.

75. "The Continuous Working Week," p. 163; Kingsbury and Fairchild, *Factory Family and Woman*, p. 248.

76. See, for example, Zhukov and Shkurat, *Seven Hours*, pp. 53–56.

77. Friedman, *Russia in Transition*, p. 262; Gruliow, "Significant Russian Approval," p. 104; Parry, "The Soviet Calendar," p. 67.

78. Friedman, ibid., p. 262; Ketchum, "Russia's Changing Tide," p. 150; "Stalin on New Economic Problems," pp. 149–50; G. Yavorski, "The Party and Government Resolutions Regarding the Regulation of the Nepreryvka are Not Being Fulfilled," *Voprosy Truda*, January 1932, pp. 72–73.

79. Friedman, ibid., p. 262.

80. Gruliow, "Significant Russian Approval," p. 104; Kingsbury and Fairchild, *Factory Family and Woman*, p. 246; Parry, "The Soviet Calendar," p. 66; "Russian Experiments," p. 69.

81. Chamberlin, *The Soviet Planned Economic Order,* pp. 162–63; Parry, ibid., p. 68; Ella Winter, *Red Virtue* (New York: Harcourt, Brace, & Co., 1933), p. 171.
82. Gruliow, "Significant Russian Approval," p. 104; Parry, ibid., p. 67; "Russian Experiments," p. 69.
83. Gruliow, ibid., p. 104; Hubbard, *Soviet Labour,* p. 98; Ketchum, "Russia's Changing Tide," p. 151; Kolarz, *Religion in the Soviet Union,* p. 31; Parry, ibid., p. 68; "Russian Experiments," p. 71.
84. Gruliow, ibid., p. 104; Hubbard, ibid., pp. 95, 98; W. W. Kulski, *The Soviet Regime* (Syracuse, NY: Syracuse University Press, 1954), p. 339; A. Yugov, *Russia's Economic Front for War and Peace* (New York: Harper, 1942), p. 163.

Chapter Three

1. Weber, *Ancient Judaism,* p. 151.
2. Brian J. L. Berry, *Geography of Market Centers and Retail Distribution* (Englewood Cliffs, NJ: Prentice-Hall, 1967), p. 93.
3. Cyril S. Belshaw, *Traditional Exchange and Modern Markets* (Englewood Cliffs, NJ: Prentice-Hall, 1965), p. 55; Marvin W. Mikesell, "The Role of Tribal Markets in Morocco," *The Geographical Review* 48(1958):494–511.
4. Webster, *Rest Days,* pp. 103–5, 118–19.
5. Berry, *Geography of Market Centers,* p. 96; Skinner, "Marketing and Social Structure," pp. 12–13.
6. E. C. Hamilton Gray, *The History of Etruria* (London: Hatchards, 1868), Vol. 3, pp. 295–305; Jacques Heurgon, *Daily Life of the Etruscans* (New York: Macmillan, 1964), p. 184; Werner Keller, *The Etruscans* (New York: Knopf, 1974), p. 46; Otto-Wilhelm von Vacano, *The Etruscans in the Ancient World* (New York: St. Martin's Press, 1960), p. 19.
7. Balsdon, *Life and Leisure,* p. 59; P. Huvelin, *Essai Historique sur le Droit des Marchés et des Foires* (Paris: Arthur Rousseau, 1897), p. 87; Ovid, *Fasti* (Cambridge, MA: Harvard University Press, 1951), p. 6; Alan E. Samuel, *Greek and Roman Chronology* (Munich: C. H. Beck'sche Verlagbuchhandlung, 1972), p. 154.
8. Balsdon, ibid., p. 60; Colson, *The Week,* p. 4; W. Warde Fowler, *The Roman Festivals of the Period of the Republic* (Port Washington, NY: Kennikat Press, 1969), p. 8; Pierre Grimal, *The Civilization of Rome* (New York: Simon and Schuster, 1963), pp. 221, 481; Léon Halkin, "Le Congé des Nundines dans les Écoles Romaines," *Revue Belge de Philologie et d'Histoire* 11(1932):121–30; Huvelin, ibid., pp. 84–86, 90; Agnes K. Michels, *The Calendar of the Roman Republic* (Princeton, NJ: Princeton University Press, 1967), pp. 84–86; Webster, *Rest Days,* pp. 120–21.
9. Huvelin, ibid., pp. 97–98; Michels, ibid., p. 89.
10. James G. Carleton, "Christian Calendar," in James Hastings (ed.), *Encyclopaedia of Religion and Ethics* (New York: Charles Scribner's Sons,

1913), Vol. 3, p. 84; Degrassi, *Inscriptiones Italiae*, Vol. 13, Fasc. 2, pp. 301, 305; Schürer, "Die Siebentägige Woche," p. 40; Walter F. Snyder, "Quinto Nundinas Pompeis," *Journal of Roman Studies* 26(1936):15.

11. Degrassi, ibid., Vol. 13, Fasc. 2, p. 305.

12. Paul Bohannan, "Concepts of Time among the Tiv of Nigeria," *The Southwestern Journal of Anthropology* 9(1953):251–62; Geneviève Calame-Griaule, "L'Expression du Temps en Dogon de Sanga," in Pierre-Francis Lacroix (ed.), *L'Expression du Temps dans Quelques Langues de l'Ouest Africain* (Paris: Centre National de la Recherche Scientifique, 1972), p. 31; R. E. Dennett, *Nigerian Studies* (London: Frank Cass, 1910), pp. 77–80; Vernon G. Fagerlund and Robert H. T. Smith, "A Preliminary Map of Market Periodicities in Ghana," *The Journal of Developing Areas* 7(1970):338, 345; Jack Goody, "Time: Social Organization," in David L. Sills (ed.), *The International Encyclopaedia of the Social Sciences* (New York: Macmillan, 1968), Vol. 16, p. 34; Polly Hill, "Notes on Traditional Market Authority and Market Periodicity in West Africa," *Journal of African History* 7(1966):302–6; B. W. Hodder, "The Yoruba Rural Market," in Paul Bohannan and George Dalton (eds.), *Markets in Africa* (Evanston, IL: Northwestern University Press, 1962), pp. 107–9; B. W. Hodder, "Some Comments on Markets and Market Periodicity," in *Markets and Marketing in West Africa* (Edinburgh: Centre of African Studies, 1966), pp. 97–109; B. W. Hodder and U. I. Ukwu, *Markets in West Africa* (Ibadan, Nigeria: Ibadan University Press, 1969), pp. 96, 128; Alexis Kagame, "An Empirical Apperception of Time and the Conception of History in Bantu Thought," in *Cultures and Time* (Paris: UNESCO, 1976), p. 105; Mübeccel B. Kiray, "The Concept of Time in Rural Societies," in Frank Greenaway (ed.), *Time and the Sciences* (Paris: UNESCO, 1979), p. 134; Robert H. T. Smith, "A Note on Periodic Markets in West Africa," *African Urban Notes* 5(1970), No. 2:29–37; Robert H. T. Smith, "West African Market-Places: Temporal Periodicity and Locational Spacing," in Claude Meillassoux (ed.), *The Development of Indigenous Trade and Markets in West Africa* (London: Oxford University Press, 1971), pp. 322–24; Northcote W. Thomas, "The Week in West Africa," *Journal of the Royal Anthropological Institute of Great Britain and Ireland* 54(1924):191, 200–206; Webster, *Rest Days*, pp. 107–16, 192; Claudia Zaslavsky, *Africa Counts* (Boston: Prindle, Weber & Schmidt, 1973), p. 224.

13. Bohannan, ibid., p. 255; Hodder, "Some Comments," pp. 102–3; Hodder and Ukwu, ibid., p. 60; Thomas, ibid., pp. 196–200; Zaslavsky, ibid., p. 221.

14. Goody, "Time," p. 34; Hill, "Notes on Traditional Market Authority," p. 305; Hodder, ibid., p. 103; B. W. Hodder, "Periodic and Daily Markets in West Africa," in Claude Meillassoux (ed.), *The Development of Indigenous Trade and Markets in West Africa* (London: Oxford University Press, 1971), p. 348; Hodder and Ukwu, ibid., p. 61; Kiray, "The Concept of Time," p. 134; Webster, *Rest Days*, pp. 107–9; Zaslavsky, ibid., p. 65.

15. Hodder, "The Yoruba Rural Market," pp. 107–8.

16. Hill, "Notes on Traditional Market Authority," p. 305; Hodder, ibid., pp. 107–8.

17. See, for example, Bohannan, "Concepts of Time," p. 256.
18. See, for example, Hodder, "Some Comments," p. 108.
19. Zerubavel, *Patterns of Time*, pp. 60–83; Zerubavel, *Hidden Rhythms*, pp. 64–69.
20. Zerubavel, "The Standardization of Time," pp. 5–10.
21. R. J. Bromley, "Markets in the Developing Countries: A Review," *Geography* 56(1971):128.
22. Harvey Sacks, E. A. Schegloff, and G. Jefferson, "A Simplest Systematics for the Organization of Turn-Taking in Conversation," *Language* 50(1974):696–735.
23. Zerubavel, *Patterns of Time*, pp. 60–76.
24. Emile Durkheim, *The Division of Labor in Society* (New York: Free Press, 1964).
25. Zerubavel, *Hidden Rhythms*, pp. 67–69.
26. J. E. Esslemont, "Bahá'í Calendar, Festivals and Dates of Historic Significance," in *The Bahá'í World* (Haifa: Universal House of Justice, 1970), Vol. 13, p. 751; John Ferraby, *All Things Made New* (Wilmette, IL: Baha'i Publishing Trust, 1960), pp. 265, 279; Horace Holley, *Religion For Mankind* (London: George Ronald, 1966), p. 105.
27. Edward G. Browne, *A Traveller's Narrative* (Amsterdam: Philo, 1975), pp. 419, 424–25; Esslemont, ibid., p. 752.
28. Mírzá Ḥuseyn of Hamadan, *The New History of the Báb* (Cambridge, England: Cambridge University Press, 1893), p. xiii; J. R. Richards, *The Religion of the Baha'is* (London: Society for Promoting Christian Knowledge, 1932), p. 228.
29. See, for example, Mírzá Huseyn of Hamadan, ibid., pp. 6, 143, 157.
30. Seyyèd Ali Mohammed, *Le Béyan Persan* (Paris: Paul Geuthner, 1911).
31. Browne, *A Traveller's Narrative*, p. 421; Richards, *The Religion of the Baha'is*, pp. 228–29.
32. Browne, ibid., p. 414; Esslemont, "Bahá'í Calendar," p. 751; Ferraby, *All Things Made New*, p. 280; Holley, *Religion for Mankind*, p. 105.
33. Charles P. Bowditch, *The Numeration, Calendar Systems and Astronomical Knowledge of the Mayas* (Cambridge, England: Cambridge University Press, 1910), p. 266; Floyd G. Lounsbury, "Maya Numeration, Computation, and Calendrical Astronomy," in Charles C. Gillispie (ed.), *Dictionary of Scientific Biography* (New York: Charles Scribner's Sons, 1978), Vol. 15, Suppl. 1, pp. 760–64; Karl Menninger, *Number Words and Number Symbols* (Cambridge, MA: MIT Press, 1977), p. 404; Sylvanus G. Morley, *The Ancient Maya* (Stanford, CA: Stanford University Press, 1956, 3d edition), pp. 240–41; Eduard Seler, "The Mexican Chronology, with Special Reference to the Zapotec Calendar," in Charles P. Bowditch (transl.), *Mexican and Central American Antiquities, Calendar Systems, and History* (Washington, DC: Government Printing Office, 1904), p. 13; Demetrio Sodi, *The Great Cultures of Mesoamerica* (Mexico City: Panorama, 1983), p. 136; J. Eric S. Thompson, *Maya Hieroglyphic Writing* (Norman: University of Oklahoma Press, 1971), pp. 51–54.
34. Michael D. Coe, *The Maya* (Harmondsworth, England: Penguin, 1971), p. 183.

35. Gordon Brotherston, *Image of the New World* (London: Thames and Hudson, 1979), p. 43; Ginzel, *Handbuch*, Vol. 1, p. 442; Barbara Tedlock, *Time and the Highland Maya* (Albuquerque: University of New Mexico Press, 1982), p. 94; Thompson, *Maya Hieroglyphic Writing*, pp. 27, 151, 183.

36. Brotherston, ibid., pp. 42, 128; Antonio Lorenzo, *Calendarios Mayas* (San-Angel, Mexico: Miguel Ángel Porrúa, 1980), pp. 83–84; Morley, *The Ancient Maya*, p. 237; Herbert J. Spinden, *Ancient Civilizations of Mexico and Central America* (New York: American Museum of Natural History, 1928), pp. 112, 119; Thompson, ibid., pp. 143–48.

37. Zerubavel, *Hidden Rhythms*, p. 89.

38. Morley, *The Ancient Maya*, pp. 241–43; Thompson, *Maya Hieroglyphic Writing*, pp. 149–59, 314–16.

39. Alfonso Caso, *The Aztecs—People of the Sun* (Norman: Oklahoma University Press, 1958), p. 68; Alfonso Caso, *Los Calendarios Prehispanicos* (Mexico City: Universidad Nacional Autónoma de México, 1967), pp. 33–39; Diego Durán, *Book of the Gods and Rites and the Ancient Calendar* (Norman: Oklahoma University Press, 1971), pp. 412–67; Munro S. Edmonson, *The Ancient Future of the Itza* (Austin: University of Texas Press, 1982), p. 198; Lorenzo, *Calendarios Mayas*, pp. 39–49; Antonio Lorenzo, *Uso e Interpretacion del Calendario Azteca* (Mexico City: Miguel Ángel Porrúa, 1983), pp. 41–45; Morley, *The Ancient Maya*, pp. 222–23, 231; Bernardino de Sahagún, *Florentine Codex* (Santa Fe, NM: The School of American Research and the University of Utah, 1951), Book 2; Jacques Soustelle, *Daily Life of the Aztecs* (New York: Macmillan, 1961), p. 109; Spinden, *Ancient Civilizations*, pp. 115, 139; Thompson, ibid., pp. 104–22; George Vaillant, *Aztecs of Mexico* (Garden City, NY: Doubleday, 1962), pp. 161, 167.

40. Edward J. Payne, *History of the New World Called America* (Oxford: Clarendon, 1899), Vol. 2, p. 359; Webster, *Rest Days*, p. 119.

41. Bowditch, *The Numeration*, p. 266; Burr C. Brundage, *The Phoenix of the Western World* (Norman: Oklahoma University Press, 1982), pp. 136–43; Caso, *The Aztecs*, pp. 65–66; Caso, *Los Calendarios*, pp. 4–15, 21; Durán, *Book of the Gods*, pp. 399–406; Ginzel, *Handbuch*, Vol. 1, pp. 434–38; Kroeber, *Anthropology*, pp. 548–49; Lorenzo, *Calendarios Mayas*, pp. 55–58; Morley, *The Ancient Maya*, pp. 205–6, 230–31; Sahagún, *Florentine Codex*, Book 4; Seler, "The Mexican Chronology," p. 14; Eduard Seler, *Comentarios al Códice Borgia* (Mexico City: Fondo de Cultura Económica, 1963), Vol. 1, pp. 11–162; Vol. 2, pp. 173–236; Soustelle, *Daily Life*, pp. 110–11; Spinden, *Ancient Civilizations*, pp. 112–14, 226; Thompson, *Maya Hieroglyphic Writing*, pp. 12, 66–103; Vaillant, *Aztecs of Mexico*, pp. 154–65.

42. Miguel León-Portilla, *Time and Reality in the Thought of the Maya* (Boston: Beacon, 1973), pp. 84, 135; Morley, ibid., pp. 190, 203–4, 240; Thompson, ibid., pp. 10, 12, 99, 129, 210. See also Caso, *Los Calendarios*, pp. 19–20.

43. Caso, ibid., pp. 21, 26; Thompson, p. 214.

44. Sahagún, *Florentine Codex*, Book 4, p. 2.

45. To appreciate the significance of the thirteen-day week, note the actual organization and physical layout of those sections of the Florentine and Borgia codices that deal with the Aztec *tonalamatl* (Sahagún, ibid.; Seler, *Comentarios*, Vol. 2, pp. 173–236). See also Lorenzo, *Uso e Interpretacion*, pp. 151–70.

46. Spinden, *Ancient Civilizations*, p. 113. See also Kroeber, *Anthropology*, pp. 548–50; Morley, *The Ancient Maya*, p. 234.

47. E. Förstemann, "The Day Gods of the Mayas," in Charles P. Bowditch (transl.), *Mexican and Central American Antiquities, Calendar Systems, and History* (Washington, DC: Government Printing Office, 1904), pp. 559–72; Morley, ibid., pp. 205–6; Thompson, *Maya Hieroglyphic Writing*, pp. 88–96.

48. James B. Greenberg, *Santiago's Sword* (Berkeley: University of California Press, 1981), pp. 111–20, 125, 129, 163; Edward T. Hall, *The Dance of Life* (Garden City, NY: Doubleday Anchor, 1984), pp. 81–90; León-Portilla, *Time and Reality*, pp. 143–49, 154–58; Lounsbury, "Maya Numeration," p. 813; Suzanna W. Miles, "An Analysis of Modern Middle American Calendars: A Study in Conservation," in Sol Tax (ed.), *Acculturation in the Americas* (New York: Cooper Square, 1967), pp. 273–84; Manning Nash, "Cultural Persistences and Social Structure: The Mesoamerican Calendar Survivals," *Southwestern Journal of Anthropology* 13(1957):149–55; Tedlock, *Time and the Highland Maya*.

49. Brundage, *The Phoenix*, pp. 143–44; Morley, *The Ancient Maya*, pp. 234–36; Thompson, *Maya Hieroglyphic Writing*, pp. 123, 150–53.

50. See, for example, William E. Gates, *The Maya and Tzental Calendars* (Cleveland, OH: 1900).

51. Spinden, *Ancient Civilizations*, pp. 117–18.

52. Hubert H. Bancroft, *The Native Races of the Pacific States of North America* (San Francisco: A. L. Bancroft, 1882), p. 511; Durán, *Book of the Gods*, p. 274; Edmonson, *The Ancient Future*, p. 198; Greenberg, *Santiago's Sword*, p. 118; Sahagún, *Florentine Codex*, Book 4, pp. 138, 144; Vaillant, *Aztecs of Mexico*, p. 55; Webster, *Rest Days*, p. 118.

53. Bancroft, ibid., p. 515; Caso, *Los Calendarios*, p. 20; E. Förstemann, "Central American Tonalamatl," in Charles P. Bowditch (transl.), *Mexican and Central American Antiquities, Calendar Systems, and History* (Washington, DC: Government Printing Office, 1904), p. 527; Lounsbury, "Maya Numeration," p. 767; Morley, *The Ancient Maya*, pp. 190, 203, 243; Seler, "The Mexican Chronology," p. 18; Seler, *Comentarios*, Vol. 1, pp. 163–78; Soustelle, *Daily Life*, p. 111; Thompson, *Maya Hieroglyphic Writing*, pp. 12, 99, 210.

54. J. G. De Casparis, *Indonesian Chronology* (Leiden and Cologne: E. J. Brill, 1978), p. 5.

55. Webster, *Rest Days*, pp. 103–5.

56. De Casparis, *Indonesian Chronology*, p. 4; Rudolf Friederich, *The Civilization and Culture of Bali* (Calcutta: Susil Gupta, 1959), p. 149; Roelof Goris, "Holidays and Holy Days," in *Bali—Studies in Life, Thought, and Ritual* (The Hague, Netherlands and Bandung, Indonesia: W. Van Hoeve, 1960), p. 117.

57. Alice G. Dewey, *Peasant Marketing in Java* (New York: Free Press, 1962), pp. 63–64, 204–10; Webster, *Rest Days*, pp. 104–5.
58. De Casparis, *Indonesian Chronology*, pp. 2–4; Goris, "Holidays," p. 118.
59. Miguel Covarrubias, *Island of Bali* (New York: Alfred A. Knopf, 1956), pp. 283–85, 315; De Casparis, ibid., pp. 18, 42; Friederich, *The Civilization*, pp. 151–53; Clifford Geertz, *The Religion of Java* (Chicago and London: University of Chicago Press, 1960), pp. 31, 39; Clifford Geertz, "Person, Time, and Conduct in Bali," in his *The Interpretation of Cultures* (New York: Basic Books, 1973), pp. 393–96; Goris, ibid., pp. 120–29; Leopold E. A. Howe, "The Social Determination of Knowledge: Maurice Bloch and Balinese Time," *Man* 16(1981):224.
60. Covarrubias, ibid., pp. 283–85, 314–15; De Casparis, ibid., pp. 3–5, 25, 28, 49; Friederich, ibid., pp. 149–50, 155; Ginzel, *Handbuch*, Vol. 1, pp. 418–19; Goris, ibid., pp. 117, 120–21, 126.
61. Geertz, "Person, Time, and Conduct," p. 393; Goris, ibid., p. 120.
62. For a complete list, see Goris, ibid., pp. 120–29.
63. See, for example, Goris, ibid., pp. 122–29.
64. Geertz, "Person, Time, and Conduct," p. 393.
65. Daniel N. Maltz, "Primitive Time-reckoning as a Symbolic System," *Cornell Journal of Social Relations* 3(1968), No. 2, p. 90.
66. Kiray, "The Concept of Time," p. 135; Martin P. Nilsson, *Primitive Time-Reckoning* (Lund, Sweden: C. W. K. Gleerup, 1920), p. 327.
67. Bancroft, *The Native Races*, p. 515; Bowditch, *The Numeration*, p. 266; Caso, *The Aztecs*, p. 66; Morley, *The Ancient Maya*, p. 163; Sahagún, *Florentine Codex*, Book 2, pp. 35–39, Book 4, p. 145; Spinden, *Ancient Civilizations*, p. 238; Vaillant, *Aztecs of Mexico*, p. 90.
68. In addition to the numerous other weekly cycles that I have mentioned (even in passing) throughout this chapter and which are not seven days long, see also Paul B. du Chaillu, *The Viking Age* (New York: Scribner's Sons, 1889), Vol. 1, pp. 37–38; Nilsson, *Primitive Time-Reckoning*, pp. 328–29; Philip, *The Calendar*, pp. 28–29; and Webster, *Rest Days*, pp. 193–96, on the nine-day week of the ancient Germans and Celts and on the five-day *fimt* of the ancient Nordic peoples.

Chapter Four

1. Phillippe Gerigny, "Improve Without Upheaval," *Journal of Calendar Reform* 1(1931):19.
2. Zerubavel, *Patterns of Time*, p. 103.
3. Samuel Zovello, "Lessons from Playing Cards," *Journal of Calendar Reform* 5(1935):68–72.
4. Colson, *The Week*, p. 2.
5. Hans Schönfeld, "Research by the Churches," *Journal of Calendar Reform* 3(1933):13.
6. M. Hyamson, "The Proposed Reform of the Calendar," *Jewish Forum* 12(1929):5; Bertha M. Parker, *The Story of Our Calendar* (Washington, DC: American Council of Education, 1933), p. 26.

7. Julius A. Roth, *Timetables* (Indianapolis: Bobbs-Merrill, 1963), p. 44.

8. Charles E. Armstrong, "Statistical Adjustments for Calendar Irregularities," *Journal of Calendar Reform* 21(1951):27–32; *Economic Aspects of Calendar Reform* (Rochester, NY: The National Committee on Calendar Simplification for the United States); Walter Mitchell, "Weekly Accounting Systems," *Journal of Calendar Reform* 4(1934):26–32; H. Platzer, "Statistical Errors," *Journal of Calendar Reform* 2(1932):90–100.

9. Henry W. Bearce, "Evolution and Revolution," *Journal of Calendar Reform* 4(1934):22.

10. Schönfeld, "Research by the Churches," p. 13.

11. William T. Bawden, "Time Problems of Schools," *Journal of Calendar Reform* 1(1931):101; Eve Beard, "School Days," *Journal of Calendar Reform* 23(1953):81.

12. Zerubavel, "Easter and Passover."

13. Balsdon, *Life and Leisure*, p. 60; Colson, *The Week*, p. 4; Fowler, *The Roman Festivals*, p. 8; Huvelin, *Essai Historique*, p. 88; Michels, *The Calendar*, pp. 23, 115, 214–15; Ovid, *Fasti*, p. 6; Samuel, *Greek and Roman Chronology*, pp. 153–54; Snyder, "Quinto Nundinas Pompeis," pp. 12–16; Webster, *Rest Days*, p. 123; W. E. van Wijk, *Le Nombre d'Or* (The Hague: Martinus Nijhoff, 1936), pp. 24–25.

14. Carleton, "Christian Calendar," pp. 84, 90; Webster, ibid., p. 123.

15. William E. Addis and Thomas Arnold (eds.), *A Catholic Dictionary* (London: Routledge & Kegan Paul, 1957. Revised 16th edition), p. 242; Seabury, *The Theory and Use*, p. 33; Herbert Thurston, "Dominical Letter," in *The Catholic Encyclopedia* (New York: Encyclopedia Press, 1913), Vol. 5, pp. 109–10.

16. Van Wijk, *Le Nombre d'Or*, pp. 114–16. See also Seabury, ibid., p. 36.

17. Peter Archer, *The Christian Calendar and the Gregorian Reform* (New York: Fordham University Press, 1941), p. 3; Seabury, ibid., p. 37.

18. L. S. Lloyd, "Intervals," in Eric Blom (ed.), *Grove's Dictionary of Music and Musicians*, 5th ed. (New York: St. Martin's Press, 1954), Vol. IV, p. 523.

19. Max Weber, "The Social Psychology of the World Religions," in Hans H. Gerth and C. Wright Mills (eds.), *From Max Weber* (New York: Oxford University Press, 1958), p. 281; Max Weber, *The Rational and Social Foundations of Music* (Carbondale, IL: Southern Illinois University Press, 1969).

20. *Enoch* 74.12. See also 82.6.

21. *Jubilees* 6.30–32. Emphasis added.

22. *Enoch* 75.1–2, 82.4,11–12; *Jubilees* 6.23–28.

23. *Enoch* 72.8–32, 82.15–20.

24. *Jubilees* 6.29.

25. Godfrey R. Driver, *The Judaean Scrolls* (Oxford: Basil Blackwell, 1965), p. 322; Annie Jaubert, "Le Calendrier des Jubilés et les Jours Liturgiques de la Semaine," *Vetus Testamentum* 7(1957):38–40. See also John M. Allegro, *The Dead Sea Scrolls* (Harmondsworth, England: Penguin, 1956), p. 114; Annie Jaubert, "Le Calendrier des Jubilés et de la Secte de Qumran: Ses Origines Bibliques," *Vetus Testamentum* 3(1953):254–

55; E. R. Leach, "A Possible Method of Intercalation for the Calendar of the Book of Jubilees," *Vetus Testamentum* 7(1957):393; Solomon Zeitlin, "Notes Relatives au Calendrier Juif," *Revue des Études Juives* 89(1930):350–54. On Reverend George Searle's similar proposal, in 1905, to add an intercalary 7-day interval to every fifth 364-day year, see Edward S. Schwegler, "Priests and the New Plan," *Journal of Calendar Reform* 4(1934):12. See also, in this connection, Alexander Philip, *The Reform of the Calendar* (London: Kegan Paul, Trench, Trübner & Co., 1914), p. 51; Meredith N. Stiles, *The World's Work and the Calendar* (Boston: Richard G. Badger, 1933), p. 154; and Wilson, *The Romance,* p. 284.

26. Genesis 1.14.
27. Jaubert, "Le Calendrier des Jubilés et les Jours," pp. 37–38.
28. *Jubilees* 2.9.
29. *Enoch* 72–73.
30. *Jubilees* 4.17–18.
31. Genesis 5.23.
32. *Jubilees* 6.32–38. Emphasis added.
33. Morgenstern, "The Calendar." On the ubiquitous presence of precise multiples of 7, 91, and 364 within the context of biblical demographic statistics, see also M. Barnouin, "Les Recensements du Livre des Nombres et l'Astronomie Babylonienne," *Vetus Testamentum* 27(1977):280–303.
34. Driver, *The Judaean Scrolls*, pp. 230, 318–21; Louis Finkelstein, *The Pharisees* (Philadelphia: Jewish Publication Society of America, 1962), Vol. 2, pp. 641–54; Van Goudoever, *Biblical Calendars*, pp. 15–29; George F. Moore, *Judaism in the First Centuries of the Christian Era* (Cambridge, MA: Harvard University Press, 1958), Vol. 1, p. 194; Julian Obermann, "Calendaric Elements in the Dead Sea Scrolls," *Journal of Biblical Literature* 75(1956):292–93; Ellis Rivkin, *A Hidden Revolution* (Nashville, TN: Abingdon, 1978), pp. 263–67; Shmaryahu Talmon, "The Calendar Reckoning of the Judaean Desert Sect," in Yigael Yadin and Chaim Rabin (eds.), *Studies in the Dead Sea Scrolls* (Jerusalem: Heichal Hasefer, 1961), pp. 78–79; Solomon Zeitlin, "The Book of Jubilees: Its Character and Its Significance," *Jewish Quarterly Review* (new series) 30(1939):14; Zerubavel, *Hidden Rhythms*, pp. 75–76, 79–81.
35. *Jubilees* 50.6–13.
36. Zerubavel, "Easter and Passover," p. 287.
37. Allegro, *The Dead Sea Scrolls*, p. 117; Millar Burrows, *The Dead Sea Scrolls* (New York: Viking, 1955), pp. 238–42; Millar Burrows, *More Light on the Dead Sea Scrolls* (New York: Viking, 1958), p. 375; Driver, *The Judaean Scrolls*, pp. 316–30; Charles T. Fritsch, *The Qumran Community* (New York: Macmillan, 1956), p. 70; Van Goudoever, *Biblical Calendars*, pp. 62–70; Jaubert, "Le Calendrier des Jubilés et de la Secte," p. 251; Jaubert, "Le Calendrier des Jubilés et les Jours," pp. 38, 61; A. R. C. Leaney, *The Rule of Qumran and Its Meaning* (Philadelphia: Westminster, 1966), pp. 80–107; Lucetta Mowry, *The Dead Sea Scrolls and the Early Church* (Chicago and London: University of Chicago Press, 1962), pp. 207–10; Obermann, "Calendaric Elements"; J. van der Ploeg, *The*

Excavations at Qumran (London: Longman, Green & Co., 1958), pp. 126–30; Chaim Rabin, *Qumran Studies* (London: Oxford University Press, 1957), pp. 77–81; Kurt Schubert, *The Dead Sea Community* (Westport, CT: Greenwood, 1959), pp. 57–58; Edmund Sutcliffe, *The Monks of Qumran* (Westminster, MD: Newman, 1960), pp. 112–13; Talmon, "The Calendar Reckoning," pp. 83–89; Geza Vermes, *The Dead Sea Scrolls* (Cleveland: Collins & World, 1978), pp. 175–78; Yigael Yadin (ed.), *The Scroll of the War of the Sons of Light Against the Sons of Darkness* (Oxford: Oxford University Press, 1962), pp. 202–6, 262.

38. Otto Neugebauer, *Ethiopic Astronomy and Computus* (Vienna: Österreichischen Akademie der Wissenschaften, 1979).

39. Mitchell, "Weekly Accounting Systems."

40. Elisabeth Achelis, *The Calendar for Everybody* (New York: G. P. Putnam's Sons, 1943), pp. 77–78.

41. Durán, *Book of the Gods*, pp. 395, 469–70; Sahagún, *Florentine Codex*, Book 2, pp. 35, 157–58, Book 4, p. 137; Seler, "The Mexican Chronology," pp. 16–17; Soustelle, *Daily Life*, p. 109; Thompson, *Maya Hieroglyphic Writing*, p. 106; Webster, *Rest Days*, pp. 280–81.

42. Achelis, *The Calendar*, pp. 53–54; Ginzel, *Handbuch*, Vol. 3, p. 350; Philip, *The Reform*, p. 56; Schwegler, "Priests and the New Plan," pp. 9–11; Wilson, *The Romance*, pp. 257, 334.

43. Ginzel, ibid., Vol. 3, p. 350; Philip, ibid., p. 56; Schwegler, ibid., p. 10.

44. C. David Stelling, "From the House of Commons," *Journal of Calendar Reform* 1(1931):115.

45. Bristow Adams, "Popular Acceptance," *Journal of Calendar Reform* 1(1931):4.

46. Webster, *Rest Days*, pp. 281–82.

47. Guillaume, *Procès-Verbaux*, Vol. 2, pp. 439, 582, 704–6.

48. "The Continuous Working Week," pp. 168–69; Echlin, "How All Nations Agree," pp. 26–27; Gruliow, "Significant Russian Approval," p. 104; Ketchum, "Russia's Changing Tide," pp. 149–50; Kingsbury and Fairchild, *Factory Family and Woman*, p. 247; Parry, "The Soviet Calendar," pp. 63, 66.

49. Guillaume, *Procès-Verbaux*, Vol. 2, pp. 444, 448, 701–2, 883.

50. Webster, *Rest Days*, p. 108.

51. Robert Graves, *The White Goddess* (New York: Farrar, Straus & Giroux, 1966), pp. 94–95, 166–67, 249.

52. Stiles, *The World's Work*, pp. 68–69.

53. Auguste Comte, *Calendrier Positiviste* (Paris: Librairie Scientifique-Industrielle de Mathias, 1852. 4th edition); Gertrud Lenzer (ed.), *Auguste Comte and Positivism* (New York: Harper Torchbooks, 1975), pp. 466–76.

54. Moses B. Cotsworth, *The Evolution of Calendars and How to Improve Them* (Washington, DC: Government Printing Office, 1922); George Eastman, *Report of the National Committee on Calendar Simplification for the United States* (Rochester, NY: 1929).

55. Stiles, *The World's Work*, pp. 57–58. See also Comte, *Calendrier Positiv-*

iste, pp. 3–5; J.-F. Marquet, "Les Calendriers Positivistes," *Annales de Bretagne et des Pays de l'Ouest* 83 (1976):373.

56. Achelis, *The Calendar*, p. 62; J. H. Hertz, *The Battle for the Sabbath at Geneva* (London: Humphrey Milford and Oxford University Press, 1932); *Journal of Calendar Reform* 16(1946):99–122, 17(1947):12–17, 99–104, 19(1949):143–45; Philip, *The Reform*, pp. 55–61; Wilson, *The Romance*, pp. 335–36.

57. Eastman, *Report of the National Committee*, pp. 8–9, 83–97.

58. Stiles, *The World's Work*, pp. 91–99.

59. *Journal of Calendar Reform* 16(1946):9–12, 17(1947):81–90, 131–37.

60. Hertz, *The Battle*; Hyamson, "The Proposed Reform"; Moses Jung, "The Opposition to the Thirteen Months Calendar," *Jewish Forum* 13(1930): 421–28; Millgram, *Sabbath*, p. 357; Isaac Rosengarten, "Religious Freedom and Calendar Reform," *Jewish Forum* 13(1930):5–7; Lawrence Wright, *Clockwork Man* (London: Elek, 1968), p. 195.

61. Hyamson, ibid., p. 7; Rosengarten, ibid., pp. 6–7.

62. Hertz, *The Battle*, pp. 9, 34. See also Rosengarten, ibid., p. 7.

63. Hertz, ibid., pp. 12, 24–26, 52; Hyamson, "The Proposed Reform," p. 6; Jung, "The Opposition," pp. 422, 425; Millgram, *Sabbath*, p. 357; Rosengarten, ibid., p. 5.

64. Hertz, ibid., pp. 27, 52; Millgram, ibid., p. 357; Rosengarten, ibid., p. 5.

65. Hertz, ibid., p. 57; Gordon Moyer, "The Gregorian Calendar," *Scientific American* 246(1982), No. 5, p. 147.

66. Seabury, *The Theory and Use*, pp. 217–18.

Chapter Five

1. Pitirim A. Sorokin, *Social and Cultural Dynamics* (New York: Bedminster, 1937), Vol. 1, pp. 161–73.

2. See, for example, Howe, "The Social Determination of Knowledge."

3. Corinne D. Bliss, *The Same River Twice* (New York: Atheneum, 1982), p. 11.

4. Zerubavel, *Hidden Rhythms*, pp. 108–14. For Eliade's general theory about the circular conception of time, see Eliade, *Cosmos and History*; Eliade, *The Sacred and the Profane*, pp. 68–113.

5. See also Lloyd W. Warner, *The Family of God* (New Haven, CT: Yale University Press, 1961), pp. 353–61.

6. Zerubavel, *Hidden Rhythms*, pp. 31–40, 49–54.

7. Moore, *Man, Time, and Society*, pp. 25–26; Georg Simmel, *The Philosophy of Money* (London: Routledge & Kegan Paul, 1978), p. 490; E. P. Thompson, "Time, Work-Discipline, and Industrial Capitalism," *Past and Present* 38(1967):70–79; Zerubavel, ibid., pp. 11, 153–66.

8. Mumford, *Technics and Civilization*, pp. 197–98.

9. Sebastian De Grazia, *Of Time, Work, and Leisure* (New York: Anchor, 1964), pp. 298–302; David S. Landes, *Revolution in Time* (Cambridge,

MA: Harvard University Press, 1983), pp. 72–75; Mumford, ibid., pp. 12–18, 197–98, 269–71; Zerubavel, *Hidden Rhythms*, p. 41.

10. Zerubavel, "The Standardization of Time," p. 10.

11. Zerubavel, *Patterns of Time*, pp. 40–42.

12. Ibid., pp. 16–17.

13. "Where Weekend Stays Are Your Worst Bet," *Business Week*, August 8, 1977, p. 74.

14. Joshua Lederberg, "Demographic Studies Related to Pediatric and Genetic Problems," unpublished paper, 1963; Paul J. Placek, K. G. Keppel, and S. M. Taffel, "Maternal Characteristics and Health Complications Associated with Cesarean Section Deliveries: Preliminary Findings from the 1980 National Fetal Mortality Survey," unpublished paper presented at the 110th annual meeting of the American Public Health Association, Montreal, November 17, 1982.

15. George W. Bohlander, *Flexitime—A New Face on the Work Clock* (Los Angeles: UCLA Institute of Industrial Relations, 1977), p. 5; Riva Poor (ed.), *4 Days 40 Hours* (New York: Mentor, 1973).

16. Douglas L. Fleuter, *The Workweek Revolution* (Reading, MA: Addison-Wesley, 1975), pp. 75–76; Albert S. Glickman and Zenia H. Brown, *Changing Schedules of Work* (Kalamazoo, MI: W. E. Upjohn Institute for Employment Research, 1974), p. 34.

17. Zerubavel, *Patterns of Time*, pp. 100–101.

18. On "counting cycles," see ibid., pp. 98–102.

19. Durkheim, *The Division of Labor*.

20. Melford E. Spiro, *Kibbutz* (New York: Schocken, 1963), p. 141. For the parallel case of a utopian American commune, see John R. Hall, *The Ways Out* (Boston: Routledge & Kegan Paul, 1978), p. 55.

21. Dahn Ben-Amotz and Netiva Ben-Yehuda, *The World Dictionary of Hebrew Slang* (Jerusalem: Lewin-Epstein, 1972), p. 217.

22. For a detailed proposal along those lines, using, however, an eight-day, rather than a seven-day, week, see John W. Pearson's utopian *The 8-Day Week* (New York: Harper & Row, 1973).

23. Zerubavel, *Patterns of Time*, pp. 21, 134. See also J. B. Priestley, *Man and Time* (Garden City, NY: Doubleday, 1964), pp. 53–55.

24. Zerubavel, ibid., pp. 17–18.

25. Archibald A. Evans, *Hours of Work in Industrialised Countries* (Geneva: International Labour Office, 1975), p. 91; Fleuter, *The Workweek Revolution*, pp. 19, 26–27, 35, 39–40; Stanley D. Nollen and Virginia H. Martin, *Alternative Work Schedules—Parts 2 and 3* (New York: American Management Associations, 1978), p. 40.

26. Evans, ibid., pp. 95–98.

27. De Grazia, *Of Time, Work, and Leisure*, pp. 422, 447; Robert Rotenberg, "Fighting with Time: Intraregional Conflicts in Public Schedules in Austria," *Urban Anthropology* 8(1979):85–88; Mary Shapcott and Phillip Steadman, "Rhythms of Urban Activity," in Tommy Carlstein, D. Parkes, and N. Thrift (eds.), *Human Activity and Time Geography* (London: Edward Arnold, 1978), pp. 66, 72; Pitirim A. Sorokin and Clarence Q. Berger,

Time-Budgets of Human Behavior (Cambridge, MA: Harvard University Press, 1939), p. 43.

28. John R. Seeley, R. A. Sim, and E. W. Loosley, *Crestwood Heights* (New York: Science Editions, 1963), p. 75.

29. See, for example, F. Stuart Chapin, *Human Activity Patterns in the City* (New York: John Wiley & Sons, 1974), pp. 98–99, 119.

30. Joseph Turow, *Entertainment, Education, and the Hard Sell* (New York: Praeger, 1981).

31. Emile Durkheim, *The Elementary Forms of the Religious Life* (New York: Free Press, 1965), pp. 24, 32, 245–51, 391–92, 488.

32. Murray Melbin, "The Colonization of Time," in Tommy Carlstein, D. Parkes, and N. Thrift (eds.), *Human Activity and Time Geography* (London: Edward Arnold, 1978), p. 106.

33. Durkheim, *The Division of Labor*, pp. 257–63; Emile Durkheim, *Suicide* (New York: Free Press, 1966), pp. 198–202.

34. Chapin, *Human Activity Patterns*, pp. 98–99, 205; De Grazia, *Of Time, Work, and Leisure*, p. 422; John P. Robinson, *How Americans Use Time* (New York: Praeger, 1977), pp. 107, 139; Sorokin and Berger, *Time-Budgets*, p. 51.

35. Chapin, ibid., p. 202.

36. See, for example, Roland Barthes, *The Fashion System* (New York: Hill and Wang, 1983), p. 251.

37. Sorokin and Berger, *Time-Budgets*, p. 52.

38. De Grazia, *Of Time, Work, and Leisure*, p. 422; Seeley, Sim, and Loosley, *Crestwood Heights*, p. 75; Sorokin and Berger, ibid., p. 34; Alexander Szalai (ed.), *The Use of Time* (The Hague: Mouton, 1972), pp. 769–71.

39. National Center for Health Statistics, "Advance Report, Final Mortality Statistics, 1979," *Monthly Vital Statistics Report*, Vol. 31 (1982), No. 6, Suppl. Department of Health and Human Services (DHHS) Publication no. [Public Health Service (PHS)] 82–1120, p. 47; Eugene Rogot, R. Fabsitz, and M. Feinleib, "Daily Variation in USA Mortality," *American Journal of Epidemiology* 103(1976):204–6.

40. Szalai, *The Use of Time*, pp. 766–68.

41. David P. Phillips, "The Fluctuation of Homicides After Publicized Executions: Reply to Kobbervig, Inverarity, and Lauderdale," *American Journal of Sociology* 88(1982):165.

42. On English fairs, see Giuseppe M. Sesti, A. T. Mann, and M. Flanagan, *The Phenomenon Book of Calendars, 1979–1980* (New York: Fireside, 1978), p. 32. See also Brian Cox, *500 Things To Do in New York for Free* (New York: Stonesong, 1982), pp. 150–71 on many annual events in the life of New York City that are fixed on particular weekends rather than on precise annual dates. For a humorous spoof parodying this calendrical practice, see the cartoons in *The New Yorker*, March 29, 1982, pp. 56–57.

43. *Mishnayot: Tamid* 7.4; *Mishnayot: Kethubboth* 1.1; *The Babylonian Talmud: Baba Ḳamma* 82a; Hayyim Schauss, *The Lifetime of a Jew* (New York: Union of American Hebrew Congregations, 1950), pp. 165–66.

44. Deuteronomy 16.9.
45. Daniel 10.2–3.
46. Zerubavel, *Hidden Rhythms*, pp. 59–64.
47. See, for example, Roth, *Timetables*, pp. 5–7, 34.
48. Thomas Mann, *The Magic Mountain* (New York: Vintage, 1969), passim.
49. Zerubavel, *Patterns of Time*, pp. 88–89.
50. See, for example, Richard Hittleman, *Yoga—28-Day Exercise Book* (New York: Bantam, 1980); Nathan Pritikin, *The Pritikin Promise* (New York: Simon and Schuster, 1983).
51. See, for example, Jack Pfeifer, *How They Train—Long Distances* (Los Altos, CA: Tafnews Press, 1982).
52. James B. Gardner and J. Gerry Purdy, *Computerized Running Training Programs* (Los Altos, CA: Tafnews Press, 1970), pp. 85–86. See also Ken Doherty, *Track and Field Omnibook* (Los Altos, CA: Tafnews Press, 1980), pp. 393–98, 405–7; David E. Martin, D. Stones, and G. Joy, *The High Jump Book* (Los Altos, CA: Tafnews Press, 1982), pp. 81–101.
53. See, for example, Martin, Stones, and Joy, ibid., pp. 49–50.
54. Förstemann, "Central American Tonalamatl," pp. 527–33; Spinden, *Ancient Civilizations*, pp. 161–63; Thompson, *Maya Hieroglyphic Writing*, pp. 214, 252–56.
55. Balsdon, *Life and Leisure*, p. 61; Abel H. J. Greenidge, *Roman Public Life* (London: Macmillan, 1911), p. 257; Grimal, *The Civilization of Rome*, p. 481; Huvelin, *Essai Historique*, pp. 92–96; A. W. Lintott, "Trinundinum," *The Classical Quarterly* 15, new series (1965):281–85; Macrobius, *The Saturnalia* (New York and London: Columbia University Press, 1969), p. 111; Michels, *The Calendar*, pp. 87, 195–97.
56. Howard I. Shapiro, *The Birth Control Book* (New York: St. Martin's Press, 1977), pp. 19–23.
57. Tarvez Tucker, *Birth Control* (New Canaan, CT: Tobey, 1975), p. 60. See also Herant A. Katchadourian and Donald T. Lunde, *Biological Aspects of Human Sexuality* (New York: Holt, Rinehart and Winston, 1975), p. 137.
58. Shapiro, *The Birth Control Book*, pp. 19–23.
59. Zerubavel, *Patterns of Time*, p. 16.
60. Sorokin, *Sociocultural Causality*, p. 201. See also pp. 172, 183–84; Henri Hubert, "Etude Sommaire de la Représentation du Temps dans la Religion et la Magie," in Henri Hubert and Marcel Mauss, *Mélanges d'Histoire des Religions* (Paris: Félix Alcan & Guillaumin, 1909), pp. 199–200.
61. Alexander M. Swaab, *School Administrator's Guide to Flexible Modular Scheduling* (West Nyack, NY: Parker, 1974), pp. 52–54.
62. On the human ability to "freeze" parts of passing, "gross" time and to measure only the remainder as "net" time, see Zerubavel, *Hidden Rhythms*, pp. 62–63.
63. Zerubavel, *Patterns of Time*, pp. 98–102.
64. Sol Robinson, *Broadcast Station Operating Guide* (Blue Ridge Summit, PA: Tab Books, 1969), pp. 40–48.
65. Turow, *Entertainment*, pp. 22–28, 57–58, 90–91.

66. W. Deane Wiley and Lloyd K. Bishop, *The Flexibly Scheduled High School* (West Nyack, NY: Parker, 1968), p. 62.
67. Gardner and Purdy, *Computerized Running*, pp. 98–99.
68. Ibid., p. 105.
69. *Cosmopolitan's Super Diets and Exercise Guide* (New York: Avon, 1974), p. 193.
70. Judith Thomas, "Co-Parenting After Divorce: Issues and Opportunities in A New Status," unpublished Ph. D. dissertation, Department of Sociology, Columbia University, 1984, pp. 81–83.
71. Zerubavel, *Hidden Rhythms*, p. 53.
72. Philip Blumstein and Pepper Schwartz, *American Couples* (New York: William Morrow, 1983), pp. 207–10.
73. Daniel Sperber, "Mishmarot and Ma'amadot," in *Encyclopaedia Judaica* (Jerusalem: Keter, 1972), Vol. 12, pp. 91–92.
74. *The Rule of Saint Benedict* (London: Sheed and Ward, 1976), Chaps. 35, 38.
75. To further appreciate the difference between these two forms of division of labor, contrast also the "alternating-week" and "split-week" schedules used by different "co-parents" (Thomas, "Co-Parenting," pp. 80–83).
76. William S. Baring-Gould and Ceil Baring-Gould, *The Annotated Mother Goose* (New York: World Publishing Co., 1967), p. 220.
77. Edward K. ("Duke") Ellington, *Music Is My Mistress* (New York: Da Capo, 1976), p. 10. See also Judith Goode, K. Curtis, and J. Theophano, "Meal Formats, Meal Cycles, and Menu Negotiation in the Maintenance of an Italian-American Community," in Mary Douglas (ed.), *Food in the Social Order* (New York: Russell Sage, 1984), pp. 150–52.
78. *The Rule of Saint Benedict*, Ch. 18.
79. Fred Wilt, *How They Train, Vol. 1—Middle Distances* (Los Altos, CA: Tafnews Press, 1973), pp. 9–10.
80. Doherty, *Track and Field Omnibook*, pp. 393–98, 405–7; Gardner and Purdy, *Computerized Running*, pp. 79, 85–86, 91–100, 104; Pfeifer, *How They Train*, passim; Wilt, ibid., passim. On sprinters, hurdlers, and high-jumpers, see also Martin, Stones, and Joy, *The High-Jump Book*, pp. 81–101; Fred Wilt, *How They Train, Vol. 3—Sprinting and Hurdling* (Los Altos, CA: Tafnews Press, 1973), passim.
81. John Barth, *The Friday Book* (New York: G. P. Putnam's Sons, 1984), p. xii.

Chapter Six

1. Henri Bergson, *Time and Free Will* (New York: Harper Torchbooks, 1960), pp. 98–128, 226–40.
2. Hubert, "Etude Sommaire," pp. 197, 207–10, 226–29. See also Sorokin, *Sociocultural Causality*, pp. 181–225.
3. Schwartz, "The Social Context of Commemoration"; Lloyd W. Warner,

The Living and the Dead (New Haven, CT: Yale University Press, 1959), pp. 129–35. See also Claude Lévi-Strauss, *The Savage Mind* (Chicago: University of Chicago Press, 1966), pp. 258–61.

4. Pitirim Sorokin and Robert K. Merton, "Social Time: A Methodological and Functional Analysis," *American Journal of Sociology* 42(1937):621.

5. Mary Douglas, *Purity and Danger* (New York: Praeger, 1966), p. 64.

6. Robert H. Rimmer, *Thursday, My Love* (New York: Signet, 1973).

7. Stanford M. Lyman and Marvin B. Scott, "On the Time Track," in *A Sociology of the Absurd* (New York: Appleton-Century-Crofts, 1970), pp. 201–2.

8. John Horton, "Time and Cool People," *Trans-Action* 4(1967), No. 5, p. 8; Dorothy Nelkin, "Unpredictability and Life Style in a Migrant Labor Camp," *Social Problems* 17(1970):479.

9. Loren Reid, *Hurry Home Wednesday* (Columbia: University of Missouri Press, 1978), p. 1.

10. Andrew J. Weigert, *Sociology of Everyday Life* (New York and London: Longman, 1981), p. 210.

11. Charlotte Brontë, *Shirley* (New York: Harper, 1900), p. 1.

12. Vid Pecjak, "Verbal Synesthesiae of Colors, Emotions, and Days of the Week," *Journal of Verbal Learning and Verbal Behavior* 6(1970):625.

13. Jim Davis, *Garfield Sits Around the House* (New York: Ballantine, 1983). In many of Davis's cartoons, Garfield is presented as the prototypical Monday-hater.

14. "Monday, Monday," by John Phillips (Duchess Music Publishers, 1965).

15. Cecil Gordon, A. R. Emerson, and D. S. Pugh, "Patterns of Sickness Absence in A Railway Population," *British Journal of Industrial Medicine* 16(1959):237–38; P. J. Taylor, "Shift and Day Work: a Comparison of Sickness Absence, Lateness, and Other Absence Behavior at an Oil Refinery from 1962 to 1965," *British Journal of Industrial Medicine* 24(1967):93–102; Zerubavel, *Patterns of Time*, pp. 16–17.

16. Gordon, Emerson, and Pugh, ibid., pp. 235–38.

17. Douglas A. Reid, "The Decline of Saint Monday 1766–1876," *Past and Present* 71(1976):76–101; Thompson, "Time, Work-Discipline, and Industrial Capitalism," pp. 73–76.

18. Wright, *Clockwork Man*, p. 211.

19. Simon W. Rabkin, F. A. Mathewson, and R. B. Tate, "Chronobiology of Cardiac Sudden Death in Men," *Journal of the American Medical Association* 244(1980):1357–58.

20. W. Baldamus, *The Structure of Sociological Inference* (London: Martin Robertson, 1976), p. 94; National Center for Health Statistics, "Advance Report," p. 47; Judith Norback (ed.), *The Mental Health Yearbook/Directory 1979–80* (New York: Van Nostrand Reinhold, 1979), p. 748; Rogot, Fabsitz, and Feinleib, "Daily Variation," p. 205.

21. "Sunday," by Ned Miller, Chester Cohn, Jules Stein, and Bennie Krueger (CBS Songs, 1926).

22. "A Sunday Kind of Love," by Barbara Belle, Louis Prima, Anita Leonard, and Stan Rhodes (MCA, 1946).

23. J. David Lewis and Andrew J. Weigert, "The Structures and Meanings of Social Time," *Social Forces* 60(1981):441.

24. Interestingly enough, in recent years, this expression has penetrated even the traditionally workaholic Japanese society. See Lee Smith, "Cracks in the Japanese Work Ethic," *Fortune*, May 14, 1984, p. 166.

25. Gordon, Emerson, and Pugh, "Patterns of Sickness Absence," pp. 237–38.

26. Loren Reid, *Finally It's Friday* (Columbia: University of Missouri Press, 1981), pp. 96–97; Alice S. Rossi and Peter E. Rossi, "Body Time and Social Time: Mood Patterns by Menstrual Cycle Phase and Day of the Week," *Social Science Research* 6(1977):289.

27. "Saturday Child," by D. Gates (Screen Gems—Columbia Music).

28. "Sunday in the Park," by Harold J. Rome (Mills Music Inc., 1937).

29. Baldamus, *The Structure*, p. 94; Rogot, Fabsitz, and Feinleib, "Daily Variation," p. 205.

30. Sándor Ferenczi, *Further Contributions to the Theory and Technique of Psycho-Analysis* (Vol. 11 of the International Psycho-Analytical Library, edited by Ernest Jones) (London: Hogarth, 1950), pp. 174–76.

31. De Grazia, *Of Time, Work, and Leisure*, p. 260.

32. Zerubavel, *Patterns of Time*, pp. 114, 117–23.

33. R. C. Browne, "The Day and Night Performance of Teleprinter Switchboard Operators," *Occupational Psychology* 23(1949):121–26.

34. Louise Bernikow, "Alone," *The New York Times Magazine*, August 15, 1982, p. 25.

35. Georg Simmel, *The Sociology of Georg Simmel* (New York: Free Press, 1964), pp. 126–32; Eviatar Zerubavel, "Personal Information and Social Life," *Symbolic Interaction* 5(1982), No. 1, pp. 100–102.

36. Zerubavel, *Hidden Rhythms*, pp. 143–45.

37. See, for example, Erving Goffman, *Behavior in Public Places* (New York: Free Press, 1963), pp. 103–4.

38. Reed Larson et al., "Time Alone in Daily Experience: Loneliness or Renewal?" in Letitia A. Peplau and Daniel Perlman (eds.), *Loneliness* (New York: John Wiley & Sons, 1982), pp. 47–48.

39. "Saturday Night Is the Loneliest Night of the Week," by Sammy Cahn (Barton Music Corp., 1944). See also "Another Saturday Night," by Sam Cooke (ABKCO Industries, Inc., 1963).

40. Colson, *The Week*, p. 29.

41. *Babylonian Talmud: Pesaḥim* 106a; Solomon Ganzfried (ed.), *The Code of Jewish Law* (New York: Hebrew Publishing Co., 1961), 96.15; Maimonides, *The Book of Seasons* (New Haven, CT: Yale University Press, 1961): *Sabbath* 29.3; Mark Zborowski and Elizabeth Herzog, *Life Is with People* (New York: Schocken, 1962), p. 37.

42. André Neher, "The View of Time and History in Jewish Culture," in *Cultures and Time* (Paris: UNESCO, 1976), p. 160.

43. *Mishnayot: Kethubboth* 1.1, *Megillah* 3.6, 4.1, *Ta'anit* 2.9, *Tamid* 7.4; *Babylonian Talmud: Bava Kamma* 37b, 82a, *Makkot* 5a, *Ta'anit* 27b, 29b.

44. Balsdon, *Life and Leisure*, p. 61; Snyder, "Quinto Nundinas," pp. 15–18.
45. Seeley, Sim, and Loosley, *Crestwood Heights*, p. 75.
46. Menninger, *Number Words*, p. 182.
47. Durkheim, *The Elementary Forms*, p. 347.
48. Ibid., p. 54.
49. Zerubavel, *Hidden Rhythms*, pp. 102–5, 110–37. See also Edmund R. Leach, *Culture and Communication* (Cambridge, England: Cambridge University Press, 1976), p. 83.
50. Hubert, "Etude Sommaire," pp. 201–4.
51. Edmund R. Leach, "Two Essays Concerning the Symbolic Representation of Time," in *Rethinking Anthropology* (London: Athlone, 1961), p. 126.
52. Ibid., pp. 133–34. Emphasis added.
53. *Babylonian Talmud: Shabbath* 113a–113b; Ganzfried, *The Code of Jewish Law* 86, 90.6; Maimonides, *The Book of Seasons: Sabbath* 24.1,4,12–13, 30.2; Zborowski and Herzog, *Life Is with People*, pp. 47–50, 56–57, 64; *The Zohar* (London: Soncino, 1949): *Beshalah* 47b.
54. Judith Goode, J. Theophano, and K. Curtis, "A Framework for the Analysis of Continuity and Change in Shared Sociocultural Rules for Food Use: the Italian-American Pattern," in Linda K. Brown and Kay Mussell (eds.), *Ethnic and Regional Foodways in the United States* (Knoxville: University of Tennessee Press, 1984), p. 74. See also Goode, Curtis, and Theophano, "Meal Formats," pp. 177, 183–91.
55. Maimonides, *The Book of Seasons: Sabbath* 30.8.
56. Zborowski and Herzog, *Life Is with People*, p. 60.
57. Barthes, *The Fashion System*, pp. 191–210; Marshall Sahlins, *Culture and Practical Reason* (Chicago and London: University of Chicago Press, 1976), p. 182.
58. *Mishnayot: Shabbath* 6.
59. *Babylonian Talmud: Shabbath* 113a; Ganzfried, *The Code of Jewish Law* 72.16; Maimonides, *The Book of Seasons: Sabbath* 30.3; *The Midrash of Psalms* (New Haven, CT: Yale University Press, 1959) 92.3.
60. Israel Abrahams, *Jewish Life in the Middle Ages* (New York: Atheneum, 1975), pp. 288–89; Samuel M. Segal, *The Sabbath Book* (New York: Thomas Yoseloff, 1957), p. 115; Zborowski and Herzog, *Life Is with People*, pp. 41–42.
61. See also Weigert, *Sociology of Everyday Life*, p. 208.
62. Barthes, *The Fashion System*, p. 251.
63. Zborowski and Herzog, *Life Is with People*, p. 363.
64. Mann, *The Magic Mountain*, p. 188.
65. Rimmer, *Thursday, My Love*, p. 21.
66. Sorokin and Merton, "Social Time," p. 624.
67. Zerubavel, *Patterns of Time*, pp. 5–8, 30–33. See also pp. 12–14.
68. Asher Koriat, B. Fischhoff, and O. Razel, "An Inquiry into the Process of Temporal Orientation," *Acta Psychologica* 40(1976):67.
69. "Lady Madonna," by John Lennon and Paul McCartney (Northern Songs, 1968).
70. Exodus 20.10–11, 23.12, 31.15, 34.21, 35.2; Deuteronomy 5.14.

71. Marc Eliot, *American Television* (Garden City, NY: Doubleday Anchor, 1981), pp. 127–273; Robinson, *Broadcast Station*, pp. 40–48.
72. Baring-Gould and Baring-Gould, *The Annotated Mother Goose*, p. 219. See also "Saturday Night," by P. G. Wodehouse (T. B. Harms and Francis, Day, and Hunter, 1916).
73. Ibid., pp. 219–20. See also "Sunday," by Ned Miller, Chester Cohn, Jules Stein, and Bennie Krueger (CBS Songs, 1926).
74. That is also true of some North American and African languages. See Arthur C. Ballard, "Calendric Terms of the Southern Puget Sound Salish," *Southwestern Journal of Anthropology* 6(1950):87; Bohannan, "Concepts of Time," p. 256; Zaslavsky, *Africa Counts*, p. 261.
75. Galit Hasan-Rokem, *Proverbs in Israeli Folk Narratives* (Helsinki: Academia Scientiarum Fennica, 1982), pp. 75–76.
76. Pfeifer, *How They Train;* Wilt, *How They Train*, Vols. 1 and 3.
77. Zerubavel, *Patterns of Time*, p. 22.
78. Eviatar Zerubavel, "The Fine Line: Boundaries and the Social Construction of Discontinuity," in progress.
79. Zerubavel, *Patterns of Time*, p. 21.
80. Zerubavel, "The Fine Line."
81. Contrast, for example, the two schedules in Zerubavel, *Patterns of Time*, pp. 19–20.
82. Zerubavel, ibid.; Zerubavel, *Hidden Rhythms*.

Chapter Seven

1. Reid, *Hurry Home Wednesday*, p. 1.
2. Thomas C. Schelling, "On the Ecology of Micromotives," *Public Interest* 25(1971):64. Emphasis added.
3. L. Erick Kanter, "Thank God It's Thursday," in Riva Poor (ed.), *4 Days 40 Hours* (New York: Mentor, 1973).
4. Charles E. Osgood, "Probing Subjective Culture: Cross-Cultural Tool-Using," *Journal of Communication* 24(1974), No. 2. pp. 95–96.
5. Derek Howse, *Greenwich Time* (Oxford: Oxford University Press, 1980), pp. 160–63.
6. Zerubavel, "The Standardization of Time," pp. 15–16.
7. For the original account of that discovery, see Antonio Pigafetta, *Magellan's Voyage Around the World* (Cleveland: Arthur H. Clark, 1906), Vol. 2, p. 185.
8. See, for example, Judah Halevi, *The Kuzari* (New York: Schocken, 1964), 2.20.
9. See, for example, Zerubavel, *Hidden Rhythms*, pp. 126–28.
10. See, for example, *The Babylonian Talmud: Shabbath* 69b.
11. Asch, *Kiddush Ha-Shem*, p. 17; Nira Harel, *Noa's Sign Language* (Jerusalem: Keter, 1979).
12. *Megillah* 13a.
13. Defoe, *Robinson Crusoe*, p. 64; Irving A. Hallowell, "Temporal Orienta-

tion in Western Civilization and in a Pre-Literate Society," *American Anthropologist* 39(1937):658.

14. Degrassi, *Inscriptiones Italiae*, Vol. 13, Fasc. 2, pp. 301, 309; Flinders Petrie, *Palestine and Israel* (London: Society for Promoting Christian Knowledge, 1934), pp. 83–84; Snyder, "Quinto Nundinas Pompeis," p. 15.

15. Philostratus, *The Life of Apollonius of Tyana* (London: William Heinemann, 1912), Vol. 1, Book 3, Ch. 41.

16. Colson, *The Week*, p. 64.

17. Zerubavel, *Hidden Rhythms*, pp. 12–30.

18. William James, *The Principles of Psychology* (New York: Dover, 1950), Vol. 1, p. 623.

19. Hans Christian Andersen, "The Roses and the Sparrows," in *Andersen's Fairy Tales* (New York: Grosset and Dunlap, 1978), p. 293.

20. Dahn Ben-Amotz, *Screwing Isn't Everything* (Tel-Aviv: Metziuth, 1979), p. 139.

21. Marcel Proust, *Swann's Way* (New York: The Modern Library, 1928), p. 139.

22. Zborowski and Herzog, *Life Is with People*, p. 363. See also p. 61.

23. Kevin Lynch, *What Time Is This Place?* (Cambridge, MA: MIT Press, 1972), p. 147.

24. Chaim N. Bialik, "The Short Friday," in Nathan Ausubel (ed.), *A Treasury of Jewish Humor* (Garden City, NY: Doubleday, 1956), pp. 59–73. See also Zerubavel, *Hidden Rhythms*, pp. 27–30.

25. Bialik, ibid., p. 70.

26. See also Asch, *Kiddush Ha-Shem*, p. 13.

27. *Shabbath* 69b.

28. Hallowell, "Temporal Orientation," pp. 650–51.

29. Sorokin, *Sociocultural Causality*, p. 193. See also p. 200.

30. Asher Koriat and Baruch Fischhoff, "What Day Is Today? An Inquiry into the Process of Time Orientation," *Memory and Cognition* 2(1974):201–5; Koriat, Fischhoff, and Razel, "An Inquiry into the Process"; Benny Shanon, "Yesterday, Today and Tomorrow," *Acta Psychologica* 43(1979):469–76.

31. Marie Jahoda, P. Lazarsfeld, and H. Zeisel, *Marienthal* (Chicago and New York: Aldine and Atherton, 1971), p. 77.

32. See my earlier discussions of the astrological seven-day week and the divinatory calendars of Indonesia and Central America. See also Balsdon, *Life and Leisure*, p. 65; A. B. Ellis, *The Tshi-Speaking Peoples of the Gold Coast of West Africa* (London: Chapman and Hall, 1887), p. 218; Ginzberg, *The Legends of the Jews*, Vol. 5, p. 39; Trachtenberg, *Jewish Magic*, pp. 254–55; Webster, *Rest Days*, pp. 127, 137, 201, 204, 222, 272–73, 290, 295.

33. Lévi-Strauss, *The Savage Mind*, p. 42.

34. *The Babylonian Talmud: Shabbath* 156a.

35. David DeCamp, "African Day-Names in Jamaica," *Language* 43(1967): 139–49; J. L. Dillard, "The West African Day-Names in Nova-Scotia,"

Names 19(1971):257–61; Ellis, *The Tshi-Speaking Peoples*, p. 219; Webster, *Rest Days*, p. 115.

36. Sorokin, *Social and Cultural Dynamics*, Vol. 4, p. 548. See also Hubert, "Etude Sommaire," p. 214; Sorokin and Merton, "Social Time," p. 624; Sorokin, *Sociocultural Causality*, p. 191.

37. Millgram, *Sabbath*, p. 337.

38. Emile Durkheim, *The Rules of Sociological Method* (New York: Free Press, 1982), pp. 50–59.

39. Berger and Luckmann, *The Social Construction*, p. 89. See also pp. 134–35; Georg Lukács, *History and Class Consciousness* (Cambridge, MA: MIT Press, 1971), pp. 83–92; Karl Marx, *Capital* (New York: International Publishers, 1967), Vol. 1, pp. 71–83.

40. Peter L. Berger, *The Sacred Canopy* (Garden City, NY: Doubleday, 1967), p. 36.

41. Peter S. Beagle, *The Last Unicorn* (New York: Ballantine, 1968), p. 199.

Bibliography

ABRAHAMS, ISRAEL. *Jewish Life in the Middle Ages.* New York: Atheneum, 1975.

ACHELIS, ELISABETH. *The Calendar for Everybody.* New York: G. P. Putnam's Sons, 1943.

ADAMS, BRISTOW. "Popular Acceptance." *Journal of Calendar Reform* 1(1931):1–6.

ADDIS, WILLIAM E., AND THOMAS ARNOLD (eds.). *A Catholic Dictionary.* London: Routledge & Kegan Paul, 1957 (revised 16th edition).

ALLEGRO, JOHN M. *The Dead Sea Scrolls.* Harmondsworth, England: Penguin, 1956.

ANDERSEN, HANS C. "The Roses and the Sparrows." Pp. 290–302 in his *Andersen's Fairy Tales.* New York: Grosset and Dunlap, 1978.

ANDREWS, GEORGE B. "Making the Revolutionary Calendar." *American Historical Review* 36(1931):515–32.

ARCHER, PETER. *The Christian Calendar and the Gregorian Reform.* New York: Fordham University Press, 1941.

ARMSTRONG, CHARLES E. "Statistical Adjustments for Calendar Irregularities." *Journal of Calendar Reform* 21(1951):27–32.

ASCH, SHOLEM. *Kiddush Ha-Shem.* Tel-Aviv: Dvir, 1953. (In Hebrew)

ATHOLL, KATHARINE (Duchess of). *The Conscription of A People.* New York: Columbia University Press, 1931.

AULARD, A. *The French Revolution.* New York: Russell & Russell, 1965.

———. *Christianity and the French Revolution.* New York: Howard Fertig, 1966.

Babylonian Talmud, The. London: Soncino, 1938.

BALDAMUS, W. *The Structure of Sociological Inference.* London: Martin Robertson, 1976.

BALLARD, ARTHUR C. "Calendric Terms of the Southern Puget Sound Salish." *Southwestern Journal of Anthropology* 6(1950):79–99.

BALSDON, J. P. V. D. *Life and Leisure in Ancient Rome.* New York: McGraw-Hill, 1969.

BANCROFT, HUBERT H. *The Native Races of the Pacific States of North America.* San Francisco: A. L. Bancroft, 1882.

BARING-GOULD, WILLIAM S., AND CEIL BARING-GOULD. *The Annotated Mother Goose.* New York: World Publishing Co., 1967.

BARNOUIN, M. "Les Recensements du Livre des Nombres et l'Astronomie Babylonienne." *Vetus Testamentum* 27(1977):280–303. (In French)

BARTH, JOHN. *The Friday Book.* New York: G. P. Putnam's Sons, 1984.

BARTHES, ROLAND. *The Fashion System.* New York: Hill and Wang, 1983.

BARTON, SAMUEL G. "The Quaker Calendar." *Proceedings of the American Philosophical Society* 93(1949):32–39.

BAWDEN, William T. "Time Problems of Schools." *Journal of Calendar Reform* 1(1931):101–2.

BEAGLE, PETER S. *The Last Unicorn.* New York: Ballantine, 1968.

BEARCE, HENRY W. "Evolution and Revolution." *Journal of Calendar Reform* 4(1934):21–25.

BEARD, EVE. "School Days." *Journal of Calendar Reform* 23(1953):81–86.

BELSHAW, CYRIL S. *Traditional Exchange and Modern Markets.* Englewood Cliffs, NJ: Prentice-Hall, 1965.

BEN-AMOTZ, DAHN. *Screwing Isn't Everything.* Tel-Aviv: Metziuth, 1979. (In Hebrew)

———, AND NETIVA BEN-YEHUDA. *The World Dictionary of Hebrew Slang.* Jerusalem: Lewin-Epstein, 1972. (In Hebrew)

BERGER, PETER L. *The Sacred Canopy.* Garden City, NY: Doubleday, 1967.

———, AND THOMAS LUCKMANN. *The Social Construction of Reality.* Garden City, NY: Doubleday Anchor, 1967.

BERGSON, HENRI. *Time and Free Will.* New York: Harper Torchbooks, 1960.

BERNIKOW, LOUISE. "Alone." *The New York Times Magazine,* August 15, 1982, pp. 24–34.

BERRY, BRIAN J. L. *Geography of Market Centers and Retail Distribution.* Englewood Cliffs, NJ: Prentice-Hall, 1967.

BIALIK, CHAIM N. "The Short Friday." Pp. 59–73 in Nathan Ausubel (ed.), *A Treasury of Jewish Humor.* Garden City, NY: Doubleday, 1956.

BLISS, CORINNE D. *The Same River Twice.* New York: Atheneum, 1982.

BLUMSTEIN, PHILIP, AND PEPPER SCHWARTZ. *American Couples.* New York: William Morrow, 1983.

BOHANNAN, PAUL. "Concepts of Time among the Tiv of Nigeria." *Southwestern Journal of Anthropology* 9(1953):251–62.

BOHLANDER, GEORGE W. *Flexitime—a New Face on the Work Clock.* Los Angeles: UCLA Institute of Industrial Relations, 1977.

BOIS, BENJAMIN. *Les Fêtes Révolutionnaires à Angers 1793–1799.* Paris: Félix Alcan, 1929. (In French)

BOLL, F. "Hebdomas." Pp. 2547–78 in Vol. 7 of *Pauly-Wissowa Real-Encyclopädie der Classischen Altertumswissenschaft.* Stuttgart: J. B. Metzler, 1912. (In German)

BOORSTIN, DANIEL J. *The Discoverers.* New York: Random House, 1983.

BOUCHÉ-LECLERCQ, AUGUSTE. *L'Astrologie Grecque.* Brussels: Culture et Civilisation, 1963. (In French)

BOWDITCH, CHARLES P. *The Numeration, Calendar Systems and Astronomical Knowledge of the Mayas.* Cambridge, England: Cambridge University Press, 1910.

BROMLEY, R. J. "Markets in the Developing Countries: a Review." *Geography* 56(1971):124–32.

BRONTË, CHARLOTTE. *Shirley.* New York: Harper, 1900.

BROTHERSTON, GORDON. *Image of the New World.* London: Thames and Hudson, 1979.

BROWNE, EDWARD G. *A Traveller's Narrative.* Amsterdam: Philo, 1975.

BROWNE, R. C. "The Day and Night Performance of Teleprinter Switchboard Operators." *Occupational Psychology* 23(1949):121–26.

BRUNDAGE, BURR C. *The Phoenix of the Western World.* Norman: University of Oklahoma Press, 1982.

BURNABY, SHERRARD B. *Elements of the Jewish and Mohammedan Calendars.* London: George Bell, 1901.

BURROWS, MILLAR. *The Dead Sea Scrolls.* New York: Viking, 1955.

———. *More Light on the Dead Sea Scrolls.* New York: Viking, 1958.

CALAME-GRIAULE, GENEVIÈVE. "L'Expression du Temps en Dogon de Sanga." Pp. 19–59 in Pierre-Francis Lacroix (ed.), *L'Expression du Temps dans Quelques Langues de l'Ouest Africain.* Paris: Centre National de la Recherche Scientifique, 1972. (In French)

CARLETON, JAMES G. "Christian Calendar." Pp. 84–91 in Vol. 3 of James Hastings (ed.), *Encyclopaedia of Religion and Ethics.* New York: Charles Scribner's Sons, 1913.

CASO, ALFONSO. *The Aztecs—People of the Sun.* Norman: University of Oklahoma Press, 1958.

———. *Los Calendarios Prehispanicos.* Mexico City: Universidad Nacional Autónoma de México, 1967. (In Spanish)

CHAMBERLIN, WILLIAM H. *The Soviet Planned Economic Order.* Boston: World Peace Foundation, 1931.

CHAPIN, F. STUART. *Human Activity Patterns in the City.* New York: John Wiley & Sons, 1974.

CLEMENT OF ALEXANDRIA. *The Stromata.* Pp. 299–567 in Vol. 2 of Alexander Roberts and James Donaldson (eds.), *The Anti-Nicene Fathers.* Grand Rapids, MI: W. B. Eerdmans, 1956.

COE, MICHAEL D. *The Maya.* Harmondsworth, England: Penguin, 1971.

COLSON, FRANCIS H. *The Week.* Cambridge, England: Cambridge University Press, 1926.

COMTE, AUGUSTE. *Calendrier Positiviste.* Paris: Librairie Scientifique-Industrielle de Mathias, 1852 (fourth edition). (In French)

"The Continuous Working Week in Soviet Russia." *International Labor Review* 23 (1931):157–80.

Cosmopolitan's Super Diets and Exercise Guide. New York: Avon, 1974.

COTSWORTH, MOSES B. *The Evolution of Calendars and How To Improve Them.* Washington, DC: Government Printing Office, 1922.

COTTON, PAUL. *From Sabbath to Sunday.* Bethlehem, PA: Times Publishing Co., 1933.

COVARRUBIAS, MIGUEL. *Island of Bali.* New York: Alfred A. Knopf, 1956.

COX, BRIAN. *500 Things To Do in New York for Free.* New York: Stonesong, 1982.

CUMONT, FRANZ. *Textes et Monuments Figurés Relatifs aux Mystères de Mithra.* Brussels: H. Lamertin, 1899. (In French)

———. *The Mysteries of Mithra.* New York: Dover, 1956.

———. *Astrology and Religion Among the Greeks and Romans.* New York: Dover, 1960.

DARESSY, GEORGES. "Une Ancienne Liste des Décans Égyptiens." *Annales du Service des Antiquités de l'Égypte* 1(1900):79–90. (In French)

DAVIS, JIM. *Garfield Sits Around the House.* New York: Ballantine, 1983.

DeCAMP, DAVID. "African Day-Names in Jamaica." *Language* 43(1967):139–49.

DE CASPARIS, J. G. *Indonesian Chronology.* Leiden and Cologne: E. J. Brill, 1978.

DEFOE, DANIEL. *The Life and Strange Surprising Adventures of Robinson Crusoe, Mariner.* London: Oxford University Press, 1972.

DEGRASSI, ATILIUS (ed.). *Inscriptiones Italiae,* Vol. 13, Fasc. 2. Rome: Libreria dello Stato, 1963. (In Latin)

DE GRAZIA, SEBASTIAN. *Of Time, Work, and Leisure.* New York: Doubleday Anchor, 1964.

DENNETT, R. E. *Nigerian Studies.* London: Frank Cass, 1910.

DEWEY, ALICE G. *Peasant Marketing in Java.* New York: Free Press of Glencoe, 1962.

DILLARD, J. L. "The West African Day-Names in Nova-Scotia." *Names* 19(1971):257–61.

DIO CASSIUS. *Dio's Roman History.* Troy, NY: Pafraets, 1905.

DOHERTY, KEN. *Track and Field Omnibook.* Los Altos, CA: Tafnews Press, 1980.

DOUGLAS, MARY. *Purity and Danger.* New York: Praeger, 1966.

———. "The Bog Irish." Pp. 59–76 in her *Natural Symbols.* New York: Vintage, 1973.

DRIVER, GODFREY R. *The Judaean Scrolls.* Oxford: Basil Blackwell, 1965.

DUBNER, P. M. "Uninterrupted Week and Labor Productivity." *Predpriyatiye* 73(1929), No. 9, pp. 45–51. (In Russian)

DU CHAILLU, PAUL B. *The Viking Age.* New York: Scribner's Sons, 1889.

DUNDAS, CHARLES. "Chagga Time Reckoning." *Man* 26(1926):140–43.

DURÁN, DIEGO. *Book of the Gods and Rites and the Ancient Calendar.* Norman: University of Oklahoma Press, 1971.

DURKHEIM, EMILE. *The Division of Labor in Society.* New York: Free Press, 1964.

———. *The Elementary Forms of the Religious Life.* New York: Free Press, 1965.

————. *Suicide.* New York: Free Press, 1966.

————. *The Rules of Sociological Method.* New York: Free Press, 1982.

EASTMAN, GEORGE. *Report of the National Committee on Calendar Simplification for the United States.* Rochester, NY, 1929.

ECHLIN, ERLAND. "How All Nations Agree." *Journal of Calendar Reform* 8(1938):25–27.

Economic Aspects of Calendar Reform. Rochester, NY: National Committee on Calendar Simplification for the United States.

EDMONSON, MUNRO S. *The Ancient Future of the Itza.* Austin: University of Texas Press, 1982.

ELIADE, MIRCEA. *Cosmos and History.* New York: Harper Torchbooks, 1959.

————. *The Sacred and the Profane.* New York: Harcourt, Brace & World, 1959.

ELIOT, MARC. *American Television.* Garden City, NY: Doubleday Anchor, 1981.

ELLINGTON, EDWARD K. ("Duke"). *Music Is My Mistress.* New York: Da Capo, 1976.

ELLIS, A. B. *The Tshi-Speaking Peoples of the Gold Coast of West Africa.* London: Chapman and Hall, 1887.

————. *The Yoruba-Speaking Peoples of the Slave Coast of West Africa.* London: Chapman and Hall, 1894.

Enoch, The Book of. Pp. 188–277 of Vol. 2 of R. H. Charles (ed.), *The Apocrypha and Pseudoepigrapha of the Old Testament.* Oxford: Oxford University Press, 1913.

Epistle of Barnabas, The. Pp. 137–49 of Vol. 1 of Alexander Roberts and James Donaldson (eds.), *The Anti-Nicene Fathers.* Grand Rapids, MI: W. B. Eerdmans, 1956.

ESSLEMONT, J. E. "Bahá'í Calendar, Festivals and Dates of Historic Significance." Pp. 749–58 of Vol. 13 of *The Bahá'í World.* Haifa: The Universal House of Justice, 1970.

EVANS, ARCHIBALD A. *Hours of Work in Industrialised Countries.* Geneva: International Labour Office, 1975.

FAGERLUND, VERNON G., AND ROBERT H. T. SMITH. "A Preliminary Map of Market Periodicities in Ghana." *The Journal of Developing Areas* 7(1970):333–47.

FERENCZI, SÁNDOR. *Further Contributions to the Theory and Technique of Psycho-Analysis.* (Vol. 11 of the International Psycho-Analytical Library, edited by Ernest Jones.) London: Hogarth Press, 1950.

FERRABY, JOHN. *All Things Made New.* Wilmette, IL: Baha'i Publishing Trust, 1960.

FINKELSTEIN, LOUIS. *The Pharisees.* Philadelphia: Jewish Publication Society of America, 1962.

FLEET, J. F. "The Use of the Planetary Names of the Days of the Week in India." *Journal of the Royal Asiatic Society,* new series, 44(1912):1039–46.

FLEUTER, DOUGLAS L. *The Workweek Revolution.* Reading, MA: Addison-Wesley, 1975.

FÖRSTEMANN, ERNST. "Central American Tonalamatl." Pp. 527–33 in Charles

P. Bowditch (transl.), *Mexican and Central American Antiquities, Calendar Systems, and History*. Washington, DC: Government Printing Office, 1904.

————. "The Day Gods of the Mayas." Pp. 559–72 in Charles P. Bowditch (transl.), *Mexican and Central American Antiquities, Calendar Systems, and History*. Washington, DC: Government Printing Office, 1904.

FOWLER, W. WARDE. *The Roman Festivals of the Period of the Republic*. Port Washington, NY: Kennikat Press, 1969.

FRIEDERICH, RUDOLF. *The Civilization and Culture of Bali*. Calcutta: Susil Gupta, 1959.

FRIEDMAN, ELISHA M. *Russia in Transition*. New York: Viking, 1932.

FRITSCH, CHARLES T. *The Qumran Community*. New York: Macmillan, 1956.

GALLERAND, L'ABBÉ J. *Les Cultes sous la Terreur en Loir-et-Cher, 1792–1795*. Paris: Grande Imprimerie de Blois, 1928. (In French)

GANDZ, SOLOMON. "The Origin of the Planetary Week or the Planetary Week in Hebrew Literature." *Proceedings of the American Academy for Jewish Research* 18(1948–49):213–54.

GANZFRIED, SOLOMON (ed.). *The Code of Jewish Law*. New York: Hebrew Publishing Co., 1961.

GARDNER, JAMES B., AND J. GERRY PURDY. *Computerized Running Training Programs*. Los Altos, CA: Tafnews Press, 1970.

GATES, WILLIAM E. *The Maya and Tzental Calendars*. Cleveland, OH: 1900.

GAXOTTE, PIERRE. *The French Revolution*. London: Charles Scribner's Sons, 1932.

GEERTZ, CLIFFORD. *The Religion of Java*. Chicago and London: University of Chicago Press, 1960.

————. "Person, Time, and Conduct in Bali." Pp. 360–411 in his *The Interpretation of Cultures*. New York: Basic Books, 1973.

GERIGNY, PHILLIPPE. "Improve Without Upheaval." *Journal of Calendar Reform* 1(1931):16–21.

GINZBERG, LOUIS. *The Legends of the Jews*. Philadelphia: The Jewish Publication Society of America, 1968.

GINZEL, F. K. *Handbuch der Mathematischen und Technischen Chronologie*. Leipzig: J. C. Hinrichs, 1906–1914. (In German)

GLICKMAN, ALBERT S., AND ZENIA H. BROWN. *Changing Schedules of Work*. Kalamazoo, MI: W. E. Upjohn Institute for Employment Research, 1974.

GOFFMAN, ERVING. *Behavior in Public Places*. New York: Free Press, 1963.

GOITEIN, S. D. "The Origin and Nature of the Muslim Friday Worship." *The Muslim World* 49(1959):183–95.

GOODE, JUDITH, K. CURTIS, AND J. THEOPHANO. "Meal Formats, Meal Cycles, and Menu Negotiation in the Maintenance of an Italian-American Community." Pp. 143–218 in Mary Douglas (ed.), *Food in the Social Order*. New York: Russell Sage, 1984.

GOODE, JUDITH, J. THEOPHANO, AND K. CURTIS. "A Framework for the Analysis of Continuity and Change in Shared Sociocultural Rules for Food Use: the Italian-American Pattern." Pp. 66–88 in Linda K. Brown and Kay Mussell (eds.), *Ethnic and Regional Foodways in the United States*. Knoxville: The University of Tennessee Press, 1984.

GOODY, JACK. "Time: Social Organization." Pp. 30–42 in Vol. 16 of David L. Sills (ed.), *The International Encyclopaedia of the Social Sciences.* New York: Macmillan, 1968.

GORDON, CECIL, A. R. EMERSON, AND D. S. PUGH. "Patterns of Sickness Absence in a Railway Population." *British Journal of Industrial Medicine* 16(1959):230–43.

GORIS, ROELOF. "Holidays and Holy Days." Pp. 115–29 in *Bali—Studies in Life, Thought and Ritual.* (Vol. 5 of the series Selected Studies on Indonesia.) The Hague, Netherlands and Bandung, Indonesia: W. Van Hoeve, 1960.

GRAVES, ROBERT. *The White Goddess.* New York: Farrar, Straus and Giroux, 1966.

GRAY E. C. HAMILTON. *The History of Etruria.* London: Hatchards, 1868.

GREENBERG, JAMES B. *Santiago's Sword.* Berkeley: University of California Press, 1981.

GREENIDGE, ABEL H. J. *Roman Public Life.* London: Macmillan, 1911.

GRÉGOIRE, HENRI. *Histoire de Sectes Religieuses.* Paris: Baudouin Frères, 1828. (In French)

GREGORY, STANFORD W. "A Quantitative Analysis of Temporal Symmetry in Microsocial Relations." *American Sociological Review* 48(1983):129–35.

GRIMAL, PIERRE. *The Civilization of Rome.* New York: Simon and Schuster, 1963.

GRIMM, JACOB. *Teutonic Mythology.* New York: Dover, 1966.

GRULIOW, LEO. "Significant Russian Approval." *Journal of Calendar Reform* 23(1953):101–5.

GUÉRIN, DANIEL. *Class Struggle in the First French Republic.* London: Pluto, 1977.

GUILLAUME, JAMES (ed.). *Procès-Verbaux du Comité d'Instruction Publique de la Convention Nationale.* Paris: Imprimerie Nationale, 1894. (In French)

HALKIN, LÉON. "Le Congé des Nundines dans les Écoles Romaines." *Revue Belge de Philologie et d'Histoire* 11(1932):121–30. (In French)

HALL, EDWARD T. *The Dance of Life.* Garden City, NY: Doubleday Anchor, 1984.

HALL, JOHN R. *The Ways Out.* Boston: Routledge & Kegan Paul, 1978.

HALLOWELL, IRVING A. "Temporal Orientation in Western Civilization and in A Pre-Literate Society." *American Anthropologist* 39(1937):647–70.

HARE, JULIUS C. "On the Names of the Days of the Week." *The Philological Museum* 1(1832):1–73.

HAREL, NIRA. *Noa's Sign Language.* Jerusalem: Keter, 1979. (In Hebrew)

HASAN-ROKEM, GALIT. *Proverbs in Israeli Folk Narratives.* Helsinki: Academia Scientiarum Fennica, 1982.

HEFELE, CHARLES J. *A History of the Councils of the Church.* Edinburgh: T. & T. Clark, 1896.

HEHN, JOHANNES. *Siebenzahl und Sabbat bei den Babyloniern und im Alten Testament.* Leipzig: Hinrichs, 1907. (In German)

HEIDEL, WILLIAM A. *The Day of Yahweh.* New York: The American Historical Association, 1929.

HERTZ, J. H. *The Battle for the Sabbath at Geneva.* London: Humphrey Milford and Oxford University Press, 1932.

HEURGON, JACQUES. *Daily Life of the Etruscans.* New York: Macmillan, 1964.

HILL, POLLY. "Notes on Traditional Market Authority and Market Periodicity in West Africa." *Journal of African History* 7(1966):295–311.

HITTLEMAN, RICHARD. *Yoga—28-Day Exercise Book.* New York: Bantam, 1980.

HODDER, B. W. "The Yoruba Rural Market." Pp. 103–17 in Paul Bohannan and George Dalton (eds.), *Markets in West Africa.* Evanston, IL: Northwestern University Press, 1962.

————. "Some Comments on Markets and Market Periodicity." Pp. 97–109 in *Markets and Marketing in West Africa.* Edinburgh: Centre of African Studies, 1966.

————. "Periodic and Daily Markets in West Africa." Pp. 347–58 in Claude Meillassoux (ed.), *The Development of Indigenous Trade and Markets in West Africa.* London: Oxford University Press, 1971.

————, AND U. I. UKWU. *Markets in West Africa.* Ibadan, Nigeria: Ibadan University Press, 1969.

HOLLEY, HORACE. *Religion for Mankind.* London: George Ronald, 1966.

Holy Scriptures, The (The Old Testament). Philadelphia: Jewish Publication Society of America, 1955.

HOOVER, CALVIN B. *The Economic Life of Soviet Russia.* New York: Macmillan, 1931.

HORTON, JOHN. "Time and Cool People." *Trans-Action* 4(1967), No. 5, pp. 5–12.

HOWE, LEOPOLD E. A. "The Social Determination of Knowledge: Maurice Bloch and Balinese Time." *Man* 16(1981):220–34.

HOWSE, DEREK. *Greenwich Time.* Oxford: Oxford University Press, 1980.

HUBBARD, LEONARD E. *Soviet Labour and Industry.* London: Macmillan, 1942.

HUBERT, HENRI. "Etude Sommaire de la Représentation du Temps dans la Religion et al Magie." Pp. 189–229 in Henri Hubert and Marcel Mauss, *Mélanges d'Histoire des Religions.* Paris: Félix Alcan and Guillaumin, 1909. (In French)

HUVELIN, P. *Essai Historique sur le Droit des Marchés et des Foires.* Paris: Arthur Rousseau, 1897. (In French)

HYAMSON, M. "The Proposed Reform of the Calendar." *Jewish Forum* 12(1929):5–7.

ICHHEISER, GUSTAV. *Appearances and Realities.* San Francisco: Jossey-Bass, 1970.

IGNATIUS. *Epistle to the Magnesians.* Pp. 59–65 in Vol. 1 of Alexander Roberts and James Donaldson (eds.), *The Anti-Nicene Fathers.* Grand Rapids, MI: W. B. Eerdmans, 1956.

ISAAC, EPHRAIM. *A New Test-Critical Introduction to Maṣḥafa Berhān.* Leiden: E. J. Brill, 1973.

JAHODA, MARIE, P. LAZARSFELD, AND H. ZEISEL. *Marienthal.* Chicago and New York: Aldine and Atherton, 1971.

JAMES, WILLIAM. *The Principles of Psychology.* New York: Dover, 1950.

JASTROW, MORRIS. *Hebrew and Babylonian Traditions.* New York: Scribner's Sons, 1914.

JAUBERT, ANNIE. "Le Calendrier des Jubilés et de la Secte de Qumran—Ses Origines Bibliques." *Vetus Testamentum* 3(1953):250–64. (In French)

――. "Le Calendrier des Jubilés et les Jours Liturgiques de la Semaine." *Vetus Testamentum* 7(1957):35–61. (In French)

JOSEPHUS. *Complete Works.* Grand Rapids, MI: Kregel, 1960.

Jubilees, The Book of. Pp. 11–82 in Vol. 2 of R. H. Charles (ed.), *The Apocrypha and Pseudoepigrapha of the Old Testament.* Oxford: Oxford University Press, 1913.

JUDAH HALEVI. *The Kuzari.* New York: Schocken, 1964.

JUNG, MOSES. "The Opposition to the Thirteen Months Calendar." *Jewish Forum* 13(1930):421–28.

JUSTIN MARTYR. *The First Apology.* Oxford: J. H. and Jas. Parker, 1861.

――. *Dialogue with Trypho.* Pp. 194–270 in Vol. 1 of Alexander Roberts and James Donaldson (eds.), *The Anti-Nicene Fathers.* Grand Rapids, MI: W. B. Eerdmans, 1956.

KAGAME, ALEXIS. "The Empirical Apperception of Time and the Conception of History in Bantu Thought." Pp. 89–116 in *Cultures and Time.* Paris: UNESCO, 1976.

KANE, PANDURANG V. *History of Dharmaśāstra.* Poona: Bhandarkar Oriental Research Institute, 1974.

KANTER, L. ERICK. "Thank God It's Thursday." Pp. 63–68 in Riva Poor (ed.), *4 Days 40 Hours.* New York: Mentor, 1973.

KANTOR, DAVID, AND WILLIAM LEHR. *Inside the Family.* San Francisco: Jossey-Bass, 1975.

KATCHADOURIAN, HERANT A., AND DONALD T. LUNDE. *Biological Aspects of Human Sexuality.* New York: Holt, Rinehart and Winston, 1975.

KELLER, WERNER. *The Etruscans.* New York: Knopf, 1974.

KETCHUM, CARLETON J. "Russia's Changing Tide." *Journal of Calendar Reform* 13(1943):147–55.

KINGSBURY, SUSAN M., AND MILDRED FAIRCHILD. *Factory Family and Woman in the Soviet Union.* New York: G. P. Putnam's Sons, 1935.

KIRAY, MÜBECCEL B. "The Concept of Time in Rural Societies." Pp. 129–42 in Frank Greenaway (ed.), *Time and the Sciences.* Paris: UNESCO, 1979.

KLIBANSKY, RAYMOND, E. PANOFSKY, AND F. SAXL. *Saturn and Melancholy.* London: Nelson, 1964.

KOLARZ, WALTER. *Religion in the Soviet Union.* New York: St. Martin's Press, 1961.

KORIAT, ASHER, AND BARUCH FISCHHOFF. "What Day Is Today? An Inquiry into the Process of Time Orientation." *Memory and Cognition* 2(1974): 201–5.

KORIAT, ASHER, B. FISCHHOFF, AND O. RAZEL. "An Inquiry into the Process of Temporal Orientation." *Acta Psychologica* 40(1976):57–73.

KROEBER, ALFRED L. *Anthropology.* New York: Harcourt, Brace and Co., 1948 (revised edition).

KRUPSKAYA, NADEZHDA. "Culture, Daily Life and the Continuous Week." Pp.

149–56 in Vol. 6 of *Pedagogicheskiye Sochineniya.* Moscow: APN, 1959. (In Russian)

KULSKI, W. W. *The Soviet Regime.* Syracuse, NY: Syracuse University Press, 1954.

LANDES, DAVID S. *Revolution in Time.* Cambridge, MA: Harvard University Press, 1983.

LANGDON, S. *Babylonian Menologies and the Semitic Calendars.* London: Oxford University Press, 1935.

LARSON, REED, et al. "Time Alone in Daily Experience: Loneliness or Renewal?" Pp. 40–53 in Letitia A. Peplau and Daniel Perlman (eds.), *Loneliness.* New York: John Wiley & Sons, 1982.

LATANÉ, BIBB, AND JOHN M. DARLEY. "Bystander 'Apathy.'" *American Scientist* 57(1969):244–68.

LAWTON, LANCELOT. "Labour." Pp. 586–625 in P. Malevsky-Malevitch (ed.), *Russia U.S.S.R.* New York: William Farquhar Payson, 1933.

LEACH, E. R. "A Possible Method of Intercalation for the Calendar of the Book of Jubilees." *Vetus Testamentum* 7(1957):392–97.

LEACH, EDMUND R. "Two Essays concerning the Symbolic Representation of Time." Pp. 124–36 in his *Rethinking Anthropology.* London: Athlone, 1961.

———. *Culture and Communication.* Cambridge, England: Cambridge University Press, 1976.

LEANEY, A. R. C. *The Rule of Qumran and Its Meaning.* Philadelphia: Westminster, 1966.

LECOMTE, HENRI. *Histoire des Théâtres de Paris—Le Théâtre de la Cité, 1792–1807.* Paris: H. Daragon, 1910. (In French)

LEDERBERG, JOSHUA. "Demographic Studies Related to Pediatric and Genetic Problems." Unpublished paper, 1963.

LEFEBVRE, GEORGES. *La Révolution Française.* Paris: Presses Universitaires de France, 1963. (In French)

———. *Napoleon—from 18 Brumaire to Tilsit, 1799–1807.* New York: Columbia University Press, 1969.

LENZER, GERTRUD (ed.). *Auguste Comte and Positivism.* New York: Harper Torchbooks, 1975.

LEÓN-PORTILLA, MIGUEL. *Time and Reality in the Thought of the Maya.* Boston: Beacon, 1973.

LÉVI-STRAUSS, CLAUDE. *The Savage Mind.* Chicago: University of Chicago Press, 1966.

LEWIS, J. DAVID, AND ANDREW J. WEIGERT. "The Structures and Meanings of Social Time." *Social Forces* 60(1981):432–62.

LEWY, HILDEGARD, AND JULIUS LEWY. "The Origin of the Week and the Oldest West Asiatic Calendar." *Hebrew Union College Annual* 17(1942–43):1–152.

LINTOTT, A. W. "Trinundinum." *The Classical Quarterly,* new series, 15(1965):281–85.

LIPSET, SEYMOUR M., M. A. TROW, AND J. S. COLEMAN. *Union Democracy.* Glencoe, IL: Free Press, 1956.

LLOYD, L. S. "Intervals." Pp. 519–24 in Vol. 4 of Eric Blom (ed.), *Grove's*

Dictionary of Music and Musicians. New York: St. Martin's Press, 1954 (5th edition).

LORENZO, ANTONIO. *Calendarios Mayas.* San-Angel, Mexico: Miguel Angel Porrúa, 1980. (In Spanish)

———. *Uso e Interpretacion del Calendario Azteca.* San-Angel, Mexico: Miguel Ángel Porrúa, 1983. (In Spanish)

LOUNSBURY, FLOYD G. "Maya Numeration, Computation, and Calendrical Astronomy." Pp. 759–818 in Suppl. 1 of Vol. 15 of Charles C. Gillispie (ed.), *Dictionary of Scientific Biography.* New York: Charles Scribner's Sons, 1978.

LUKÁCS, GEORG. *History and Class Consciousness.* Cambridge, MA: MIT Press, 1971.

LYMAN, STANFORD M., AND MARVIN B. SCOTT. "On the Time Track." Pp. 189–212 in their *A Sociology of the Absurd.* New York: Appleton-Century-Crofts, 1970.

LYNCH, KEVIN. *What Time Is This Place?* Cambridge, MA: MIT Press, 1972.

LYONS, MARTYN. *France Under the Directory.* Cambridge, England: Cambridge University Press, 1975.

MACQUEEN, JAMES G. *Babylon.* New York: Praeger, 1965.

MACROBIUS. *The Saturnalia.* New York and London: Columbia University Press, 1969.

MAIMONIDES. *The Book of Seasons.* New Haven, CT: Yale University Press, 1961.

MALTZ, DANIEL N. "Primitive Time-Reckoning as A Symbolic System." *Cornell Journal of Social Relations* 3(1968), No. 2, pp. 85–111.

MANN, J. "The Observance of the Sabbath and the Festivals in the First Two Centuries of the Current Era according to Philo, Josephus, the New Testament, and the Rabbinic Sources." *The Jewish Review* 4(1913–14):433–56, 498–532.

MANN, THOMAS. *The Magic Mountain.* New York: Vintage, 1969.

MARQUET, J. F. "Les Calendriers Positivistes." *Annales de Bretagne et des Pays de l'Ouest* 83(1976):371–78.

MARTIN, DAVID, D. STONES, AND G. JOY. *The High Jump Book.* Los Altos, CA: Tafnews Press, 1982.

MARX, KARL. *Capital.* New York: International Publishers, 1967.

MATHIEZ, ALBERT. *La Théophilanthropie et le Culte Décadaire, 1796–1801.* Paris, 1904.

———. *The French Revolution.* New York: Russell & Russell, 1962.

MEEK, THEOPHILE J. "The Sabbath in the Old Testament." *Journal of Biblical Literature* 33(1914):201–12.

MELBIN, MURRAY. "The Colonization of Time." Pp. 100–113 in Tommy Carlstein, D. Parkes, and N. Thrift (eds.), *Human Activity and Time Geography.* London: Edward Arnold, 1978.

MENNINGER, KARL. *Number Words and Number Symbols.* Cambridge, MA: MIT Press, 1977.

MERITT, BENJAMIN D. *The Athenian Year.* Berkeley: University of California Press, 1961.

MICHELS, AGNES K. *The Calendar of the Roman Republic.* Princeton, NJ: Princeton University Press, 1967.

Midrash of Psalms, The. New Haven, CT: Yale University Press, 1959.

Midrash Rabbah—Genesis. London: Soncino, 1951.

MIKESELL, MARVIN W. "The Role of Tribal Markets in Morocco." *The Geographical Review* 48(1958):494–511.

MILES, SUZANNA W. "An Analysis of Modern Middle American Calendars: a Study in Conservation." Pp. 273–84 in Sol Tax (ed.), *Acculturation in the Americas.* New York: Cooper Square, 1967.

MILLGRAM, ABRAHAM E. *Sabbath—the Day of Delight.* Philadelphia: The Jewish Publication Society of America, 1965.

MÍRZÁ ḤUSEYN OF HAMADAN. *The New History of the Báb.* Cambridge, England: Cambridge University Press, 1893.

Mishnah, The. London: Oxford University Press, 1958.

MITCHELL, WALTER. "Weekly Accounting Systems." *Journal of Calendar Reform* 4(1934):26–32.

MITTON, SIMON (ed.). *The Cambridge Encyclopaedia of Astronomy.* New York: Crown, 1977.

MOORE, GEORGE F. *Judaism in the First Centuries of the Christian Era.* Cambridge, MA: Harvard University Press, 1958.

MOORE, WILBERT E. *Man, Time, and Society.* New York: John Wiley & Sons, 1963.

MORGENSTERN, JULIAN. "The Calendar of the Book of Jubilees, Its Origin and Its Character." *Vetus Testamentum* 5(1955):34–76.

MORLEY, SYLVANUS. *The Ancient Maya.* Stanford, CA: Stanford University Press, 1956 (3rd edition).

MOWRY, LUCETTA. *The Dead Sea Scrolls and the Early Church.* Chicago and London: University of Chicago Press, 1962.

MOYER, GORDON. "The Gregorian Calendar." *Scientific American* 246(1982), No. 5, pp. 144–52.

MULDERS, GERARD F. W. "Intricacies of the Calendar." *Journal of Calendar Reform* 11(1941):134–38.

MUMFORD, LEWIS. *Technics and Civilization.* New York: Harbinger, 1963.

MURRAY, MARGARET A. *The Splendor That Was Egypt.* New York: Hawthorn, 1963 (revised edition).

NASH, MANNING. "Cultural Persistences and Social Structure: The Mesoamerican Calendar Survivals." *Southwestern Journal of Anthropology* 13(1957):149–55.

National Center for Health Statistics. "Advance Report, Final Mortality Statistics, 1979." *Monthly Vital Statistics Report* 31(1982), No. 6. [Suppl. Department of Health and Human Services (DHHS) Publication No. Public Health Service (PHS) 82-1120].

NEHER, ANDRÉ. "The View of Time and History in Jewish Culture." Pp. 149–67 in *Cultures and Time.* Paris: UNESCO, 1976.

NELKIN, DOROTHY. "Unpredictability and Life Style in A Migrant Labor Camp." *Social Forces* 17(1970):472–87.

NEUGEBAUER, OTTO. *The Exact Sciences in Antiquity.* New York: Harper Torchbooks, 1962.

———. *A History of Ancient Mathematical Astronomy.* New York: Springer-Verlag, 1975.

———. *Ethiopic Astronomy and Computus.* Vienna: Österreichischen Akademie der Wissenschaften, 1979.

New English Bible, The (The New Testament). New York: Oxford University Press, 1961.

NILSSON, MARTIN P. *Primitive Time-Reckoning.* Lund, Sweden: C. W. K. Gleerup, 1920.

NOLLEN, STANLEY D., AND VIRGINIA H. MARTIN. *Alternative Work Schedules—Parts 2 and 3.* New York: American Management Associations, 1978.

NORBACK, JUDITH (ed.). *The Mental Health Yearbook/Directory 1979–80.* New York: Van Nostrand Reinhold, 1979.

OBERMANN, JULIAN. "Calendaric Elements in the Dead Sea Scrolls." *Journal of Biblical Literature* 75(1956):285–97.

OSGOOD, CHARLES E. "Probing Subjective Culture: Cross-Cultural Tool-Using." *Journal of Communication* 24(1974), No. 2, pp. 82–100.

OVID. *Fasti.* Cambridge, MA: Harvard University Press, 1951.

PARKER, BERTHA M. *The Story of Our Calendar.* Washington, DC: American Council on Education, 1933.

PARRY, ALBERT. "The Soviet Calendar." *Journal of Calendar Reform* 10(1940):63–69.

PÀTTARO, GERMANO. "The Christian Conception of Time." Pp. 169–95 in *Cultures and Time.* Paris: UNESCO Press, 1976.

PAYNE, Edward J. *History of the New World Called America.* Oxford: Clarendon, 1899.

PEARSON, JOHN W. *The 8-Day Week.* New York: Harper and Row, 1973.

PECJAK, VID. "Verbal Synesthesiae of Colors, Emotions, and Days of the Week." *Journal of Verbal Learning and Verbal Behavior* 9(1970):623–26.

PETRIE, FLINDERS. *Palestine and Israel.* London: Society for Promoting Christian Knowledge, 1934.

———. *Wisdom of the Egyptians.* London: British School of Archaeology in Egypt, 1940.

PFEIFER, JACK. *How They Train—Long Distances.* Los Altos, CA: Tafnews Press, 1982.

PHILIP, ALEXANDER. *The Reform of the Calendar.* London: Kegan Paul, Trench, Trübner & Co., 1914.

———. *The Calendar.* New York and London: Macmillan and Cambridge University Press, 1921.

PHILLIPS, DAVID P. "The Fluctuation of Homicides after Publicized Executions: Reply to Kobbervig, Inverarity, and Lauderdale." *American Journal of Sociology* 88(1982):165–66.

PHILOSTRATUS. *The Life of Apollonius of Tyana.* London: William Heinemann, 1912.

PIERRE, CONSTANT. *Les Hymnes et Chansons de la Révolution.* Paris: Imprimerie Nationale. (In French)

PIGAFETTA, ANTONIO. *Magellan's Voyage Around the World.* Cleveland: Arthur H. Clark, 1906.

PLACEK, PAUL J., K. G. KEPPEL, AND S. M. TAFFEL. "Maternal Characteristics and Health Complications Associated with Cesarean Section Deliveries: Preliminary Findings from the 1980 National Natality Survey and 1980 National Fetal Mortality Survey." Unpublished paper presented at the 110th Annual Meeting of the American Public Health Association, Montreal, November 17, 1982.

PLATZER, H. "Statistical Errors." *Journal of Calendar Reform* 2(1932):90–100.

POOR, RIVA (ed.). *4 Days 40 Hours.* New York: Mentor, 1973.

PRIESTLEY, J. B. *Man and Time.* Garden City, NY: Doubleday, 1964.

PRITIKIN, NATHAN. *The Pritikin Promise.* New York: Simon and Schuster, 1983.

PROUST, MARCEL. *Swann's Way.* New York: Modern Library, 1928.

RABIN, CHAIM. *Qumran Studies.* London: Oxford University Press, 1957.

RABKIN, SIMON W., F. A. MATHEWSON, AND R. B. TATE. "Chronobiology of Cardiac Sudden Death in Men." *Journal of the American Medical Association* 244(1980):1357–58.

RAMSAY, GEORGE G. *The Histories of Tacitus.* London: John Murray, 1915.

REID, DOUGLAS A. "The Decline of Saint Monday 1766–1876." *Past and Present* 71(1976):76–101.

REID, LOREN. *Hurry Home Wednesday.* Columbia: University of Missouri Press, 1978.

———. *Finally It's Friday.* Columbia: University of Missouri Press, 1981.

RICHARDS, J. R. *The Religion of the Bahá'ís.* London: Society for Promoting Christian Knowledge, 1932.

RIMMER, ROBERT H. *Thursday, My Love.* New York: Signet, 1973.

RIVKIN, ELLIS. *A Hidden Revolution.* Nashville, TN: Abingdon, 1978.

ROBINSON, JOHN P. *How Americans Use Time.* New York: Praeger, 1977.

ROBINSON, SOL. *Broadcast Station Operating Guide.* Blue Ridge Summit, PA: Tab Books, 1969.

ROGOT, EUGENE, R. FABSITZ, AND M. FEINLEIB. "Daily Variation in USA Mortality." *American Journal of Epidemiology* 103(1976):198–211.

ROSENGARTEN, ISAAC. "Religious Freedom and Calendar Reform." *Jewish Forum* 13(1930):5–7.

ROSSI, ALICE S., AND PETER E. ROSSI. "Body Time and Social Time: Mood Patterns by Menstrual Cycle Phase and Day of the Week." *Social Science Research* 6(1977):273–308.

ROTENBERG, ROBERT. "Fighting with Time: Intraregional Conflict in Public Schedules in Austria." *Urban Anthropology* 8(1979):73–94.

ROTH, JULIUS A. *Timetables.* Indianapolis: Bobbs-Merrill, 1963.

Rule of Saint Benedict, The. London: Sheed and Ward, 1976.

"Russian Experiments." *Journal of Calendar Reform* 6(1936):69–71.

SACHS, A. "Babylonian Horoscopes." *Journal of Cuneiform Studies* 6(1952):49–75.

SACKS, HARVEY, E. A. SCHEGLOFF, AND G. JEFFERSON. "A Simplest Systematics for the Organization of Turn-Taking in Conversation." *Language* 50(1974):696–735.

SAHAGÚN, BERNARDINO. *The Ceremonies.* (Book 2 of the *Florentine Codex* or the *General History of the Things of New Spain.*) Santa Fe, NM: The School of American Research and the University of Utah, 1951.

――――. *The Soothsayers.* (Book 4 of the *Florentine Codex* or the *General History of the Things of New Spain.*) Santa Fe, NM: The School of American Research and the University of Utah, 1957.

SAHLINS, MARSHALL. *Culture and Practical Reason.* Chicago and London: University of Chicago Press, 1976.

SAMUEL, ALAN E. *Greek and Roman Chronology.* Munich: C. H. Beck'sche Verlagsbuchhandlung, 1972.

SCHAUSS, HAYYIM. *The Lifetime of A Jew.* New York: Union of American Hebrew Congregations, 1950.

SCHELLING, THOMAS C. "On the Ecology of Micromotives." *Public Interest* 25(1971):61–98.

SCHÖNFELD, HANS. "Research by the Churches." *Journal of Calendar Reform* 3(1933):10–18.

SCHUBERT, KURT. *The Dead Sea Community.* Westport, CT: Greenwood, 1959.

SCHÜRER, EMIL. "Die Siebentägige Woche im Gebrauche der Christlichen Kirche der Ersten Jahrhunderte." *Zeitschrift für die Neutestamentliche Wissenschaft* 6(1905):1–66. (In German)

SCHUTZ, ALFRED. "Making Music Together: a Study in Social Relationship." Pp. 159–78 in Vol. 2 of his *Collected Papers.* The Hague: Martinus Nijhoff, 1964.

SCHWARTZ, BARRY. "The Social Context of Commemoration: A Study in Collective Memory." *Social Forces* 61(1982):374–402.

SCHWEGLER, EDWARD S. "Priests and the New Plan." *Journal of Calendar Reform* 4(1934):7–14.

SEABURY, SAMUEL. *The Theory and Use of the Church Calendar in the Measurement and Distribution of Time.* New York: Pott, Young, and Co., 1872.

SEELEY, JOHN R., R. A. SIM, AND E. W. LOOSLEY. *Crestwood Heights.* New York: Science Editions, 1963.

SEGAL, SAMUEL M. *The Sabbath Book.* New York: Thomas Yoseloff, 1957.

SELER, EDUARD. "The Mexican Chronology, with Special Reference to the Zapotec Calendar." Pp. 13–55 in Charles P. Bowditch (transl.), *Mexican and Central American Antiquities, Calendar Systems, and History.* Washington, DC: Government Printing Office, 1904.

――――. *Comentarios al Códice Borgia.* Mexico City: Fondo de Cultura Económica, 1963. (In Spanish)

SESTI, GIUSEPPE M., A. T. MANN, AND M. FLANAGAN. *The Phenomenon Book of Calendars 1979–1980.* New York: Fireside, 1978.

SEYYÈD ALI MOHAMMED (The Báb). *Le Béyan Persan.* Paris: Paul Geuthner, 1911. (In French)

SHANON, BENNY. "Yesterday, Today and Tomorrow." *Acta Psychologica* 43(1979):469–76.

SHAPCOTT, MARY, AND PHILLIP STEADMAN. "Rhythms of Urban Activity." Pp. 49–74 in Tommy Carlstein, D. Parkes, and N. Thrift (eds.), *Human Activity and Time Geography.* London: Edward Arnold, 1978.

SHAPIRO, HOWARD I. *The Birth Control Book.* New York: St. Martin's Press, 1977.

SIMMEL, GEORG. *The Sociology of Georg Simmel.* New York: Free Press, 1964.

———. *The Philosophy of Money.* London: Routledge & Kegan Paul, 1978.

SKINNER, WILLIAM. "Marketing and Social Structure in Rural China." *Journal of Asian Studies* 24(1964–65):3–43, 195–228, 363–99.

SMITH, LEE. "Cracks in the Japanese Work Ethic." *Fortune,* May 14, 1984, pp. 162–68.

SMITH, ROBERT H. T. "A Note on Periodic Markets in West Africa." *African Urban Notes* 5(1970), No. 2, pp. 29–37.

———. "West African Market-Places: Temporal Periodicity and Locational Spacing." Pp. 319–46 in Claude Meillassoux (ed.), *The Development of Indigenous Trade and Markets in West Africa.* London: Oxford University Press, 1971.

SNYDER, WALTER F. "Quinto Nundinas Pompeis." *Journal of Roman Studies* 26(1936):12–18.

SODI, DEMETRIO. *The Great Cultures of Mesoamerica.* Mexico City: Panorama, 1983.

SOROKIN, PITIRIM A. *Social and Cultural Dynamics.* New York: Bedminster, 1937–41.

———. *Sociocultural Causality, Space, Time.* Durham, NC: Duke University Press, 1943.

———, and CLARENCE Q. BERGER. *Time-Budgets of Human Behavior.* Cambridge, MA: Harvard University Press, 1939.

———, and ROBERT K. MERTON. "Social Time: A Methodological and Functional Analysis." *American Journal of Sociology* 42(1937):615–29.

SOUSTELLE, JACQUES. *Daily Life of the Aztecs.* New York: Macmillan, 1961.

SPERBER, DANIEL. "Mishmarot and Ma'amadot." Pp. 89–92 in Vol. 12 of *Encyclopaedia Judaica.* Jerusalem: Keter, 1972.

SPINDEN, HERBERT J. *Ancient Civilizations of Mexico and Central America.* New York: American Museum of Natural History, 1928.

SPIRO, MELFORD E. *Kibbutz.* New York: Schocken, 1963.

"Stalin on New Economic Problems." *Soviet Union Review* 9(1931):146–54.

STELLING, C. DAVID. "From the House of Commons." *Journal of Calendar Reform* 1(1931):112–18.

STILES, MEREDITH N. *The World's Work and the Calendar.* Boston: Richard G. Badger, 1933.

STRUTYNSKI, UDO. "Germanic Divinities in Weekday Names." *Journal of Indo-European Studies* 3(1975):363–84.

SUTCLIFFE, EDMUND. *The Monks of Qumran.* Westminster, MD: Newman, 1960.

SWAAB, ALEXANDER M. *School Administrator's Guide to Flexible Modular Scheduling.* West Nyack, NY: Parker, 1974.

SWAMIKANNU PILLAI. *Indian Chronology.* Madras: Grant, 1911.

SYDENHAM, M. J. *The First French Republic, 1792–1804.* London: B. T. Batsford, 1974.

SZALAI, ALEXANDER (ed.). *The Use of Time.* The Hague: Mouton, 1972.

TALMON, SHMARYAHU. "Divergences in Calendar-Reckoning in Ephraim and Judah." *Vetus Testamentum* 8(1958):48–74.

————. "The Calendar Reckoning of the Judaean Desert Sect." Pp. 77–105 in Yigael Yadin and Chaim Rabin (eds.), *Studies in the Dead Sea Scrolls*. Jerusalem: Heichal Hasefer, 1961. (In Hebrew)

TAYLOR, P. J. "Shift and Day Work: a Comparison of Sickness Absence, Lateness, and Other Absence Behavior at An Oil Refinery from 1962 to 1965." *British Journal of Industrial Medicine* 24(1967):93–102.

Teachings of the Twelve Apostles, The. Pp. 15–25 in James A. Kleist (ed.), *Ancient Christian Writers*. Westminster, MD: Newman, 1948.

TEDLOCK, BARBARA. *Time and the Highland Maya*. Albuquerque: University of New Mexico Press, 1982.

TERTULLIAN. *Apology*. Pp. 17–55 in Vol. 3 of Alexander Roberts and James Donaldson (eds.), *The Anti-Nicene Fathers*. Grand Rapids, MI: W. B. Eerdmans, 1951.

THOMAS, JUDITH. "Co-Parenting after Divorce: Issues and Opportunities in a New Status." Unpublished doctoral dissertation, Department of Sociology, Columbia University, 1984.

THOMAS, NORTHCOTE W. "The Week in West Africa." *Journal of the Royal Anthropological Institute of Great Britain and Ireland* 54(1924):183–209.

THOMPSON, E. P. "Time, Work-Discipline, and Industrial Capitalism." *Past and Present* 38(1967):56–97.

THOMPSON, J. ERIC S. *Maya Hieroglyphic Writing*. Norman: University of Oklahoma Press, 1971.

THURSTON, HERBERT. "Dominical Letter." Pp. 109–10 in Vol. 5 of *The Catholic Encyclopedia*. New York: Encyclopedia Press, 1913.

TOLSTOY, LEO. "The Death of Ivan Ilych." Pp. 95–156 in his *The Death of Ivan Ilych and Other Stories*. New York: Signet Classics, 1960.

TRACHTENBERG, JOSHUA. *Jewish Magic and Superstition*. New York: Atheneum, 1970.

TRUMBO, DALTON. *Johnny Got His Gun*. New York: Bantam, 1970.

TUCKER, TARVEZ. *Birth Control*. New Canaan, CT: Tobey, 1975.

TUROW, JOSEPH. *Entertainment, Education, and the Hard Sell*. New York: Praeger, 1981.

VAILLANT, GEORGE. *Aztecs of Mexico*. Garden City, NY: Doubleday, 1962.

VAN DER PLOEG, J. *The Excavations at Qumran*. London: Longman, Green & Co., 1958.

VAN DER WAERDEN, B. L. "The Date of Invention of Babylonian Planetary Theory." *Archive for History of Exact Sciences* 5(1968):70–78.

VAN GOUDOEVER, J. *Biblical Calendars*. Leiden: E. J. Brill, 1961.

VAN WIJK, W. E. *Le Nombre d'Or*. The Hague: Martinus Nijhoff, 1936. (In French)

VERMES, GEZA. *The Dead Sea Scrolls*. Cleveland: Collins & World, 1978.

VON IHERING, RUDOLPH. *The Evolution of the Aryan*. New York: Henry Holt, 1897.

VON VACANO, OTTO-WILHELM. *The Etruscans in the Ancient World*. New York: St. Martin's Press, 1960.

WARNER, LLOYD W. *The Living and the Dead*. New Haven, CT: Yale University Press, 1959.

————. *The Family of God*. New Haven, CT: Yale University Press, 1961.

WEBER, MAX. "The Social Psychology of World Religions." Pp. 267–301 in Hans H. Gerth and C. Wright Mills (eds.), *From Max Weber*. New York: Oxford University Press, 1958.

———. *Ancient Judaism*. New York: Free Press, 1967.

———. *The Rational and Social Foundations of Music*. Carbondale: Southern Illinois University Press, 1969.

———. *Economy and Society*. Berkeley: University of California Press, 1978.

WEBSTER, HUTTON. *Rest Days*. New York: Macmillan, 1916.

WEIGERT, ANDREW J. *Sociology of Everyday Life*. New York and London: Longman, 1981.

WELSCHINGER, HENRI. *Le Théatre de la Révolution, 1789–1799*. Paris: Charavay Frères, 1880. (In French)

———. *Les Almanachs de la Révolution*. Paris: Librairie des Bibliophiles, 1884. (In French)

WILEY, W. DEANE, AND LLOYD K. BISHOP. *The Flexibly Scheduled High School*. West Nyack, NY: Parker, 1968.

WILSON, P. W. *The Romance of the Calendar*. New York: W. W. Norton, 1937.

WILT, FRED. *How They Train, Vol. 1—Middle Distances*. Los Altos, CA: Tafnews Press, 1973.

———. *How They Train, Vol. 3—Sprinting and Hurdling*. Los Altos, CA: Tafnews Press, 1973.

WINTER, ELLA. *Red Virtue*. New York: Harcourt, Brace & Co., 1933.

WRIGHT, LAWRENCE. *Clockwork Man*. London: Elek, 1968.

YADIN, YIGAEL (ed.). *The Scroll of the War of the Sons of Light Against the Sons of Darkness*. Oxford: Oxford University Press, 1962.

YAVORSKI, G. "The Party and Government Resolutions regarding the Regulation of the Nepreryvka Are Not Being Fulfilled." *Voprosy Truda* 1(1932), January, pp. 72–77. (In Russian)

YUGOV, A. *Russia's Economic Front for War and Peace*. New York: Harper, 1942.

ZASLAVSKY, CLAUDIA. *Africa Counts*. Boston: Prindle, Weber & Schmidt, 1973.

ZBOROWSKI, MARK, AND ELIZABETH HERZOG. *Life Is with People*. New York: Schocken, 1962.

ZEITLIN, SOLOMON. "Notes Relatives au Calendrier Juif." *Revue des Études Juives* 89(1930):349–59. (In French)

———. "The Book of Jubilees: Its Character and Its Significance." *Jewish Quarterly Review*, new series, 30(1939):1–31.

ZERUBAVEL, EVIATAR. *Patterns of Time in Hospital Life*. Chicago and London: University of Chicago Press, 1979.

———. "The Bureaucratization of Responsibility: the Case of Informed Consent." *Bulletin of the American Academy of Psychiatry and the Law* 8(1980):161–67.

———. *Hidden Rhythms*. Chicago and London: University of Chicago Press, 1981.

———. "Easter and Passover: on Calendars and Group Identity." *American Sociological Review* 47(1982):284–89.

———. "Personal Information and Social Life." *Symbolic Interaction* 5(1982), No. 1, pp. 97–109.

———. "The Standardization of Time: a Sociohistorical Perspective." *American Journal of Sociology* 88(1982):1–23.

———. "The Fine Line: Boundaries and the Social Construction of Discontinuity." (Work in progress)

ZHUKOV, N., AND A. SHKURAT. *Seven Hours—1927–1932.* Moscow: Profizdat, 1933. (In Russian)

Zohar, The. London: Soncino, 1949.

ZOVELLO, SAMUEL. "Lessons from Playing Cards." *Journal of Calendar Reform* 5(1935):68–72.

Author Index

Subject Index